ASE Guide to

SECONDARY

Science Education

Fourth Edition

Edited by
Indira Banner & Judith Hillier

The Association
for Science Education
Promoting Excellence in Science Teaching and Learning

Published by:
The Association for Science Education, College Lane, Hatfield, Herts AL10 9AA

© The Association for Science Education 2018

ISBN: 978-0-86357-458-0

Design and page layout: Commercial Campaigns
Executive Editor: Jane Hanrott

Printed by: Ashford Press, Gosport, Hampshire
Cover credit: iStockphoto.com/Wavebreakmedia

CONTENTS

Foreword v
About the authors vii
About ASE xv
Introduction xvii

SECTION 1: Foundations of science education

1   Aims and approaches of good science education
     Wynne Harlen 2

2   The nature of science and the nature of school science
     Keith S. Taber and Richard Brock 14

3   Educational neuroscience and the brain: some implications
     for our understanding of learning and teaching
     Derek Bell and Helen Darlington 26

4   Primary science and the implications for
     secondary science teachers
     Helen Wilson 40

SECTION 2: Students: all learning science

5   Educating students for the world: the role of emotions
     in learning science
     Brian Matthews 52

6   Planning for motivation and engagement
     Mark Hardman 66

7   Active learning
     Ed Walsh 79

8   Planning for student understanding and behaviour for learning
     Judith Hillier 89

9   Inclusive science education
     Lucy Dix, Fiona Woodhouse and Indira Banner 99

10   Equality and diversity in science education
     Clare Thomson 113

SECTION 3: Science teachers: synthesising learning

11   Creativity in teaching science
     Deb McGregor and Nicoleta Gaciu 132

12   Thinking about practical work
     Ian Abrahams and Nikolaos Fotou 150

13   Taking science teaching outside the classroom:
     using a pedagogical framework
     Melissa Glackin and Natasha Serret 161

14   Mathematics in science teaching
     Richard Needham and Mike Sands 173

15   Enhancing the teaching of science with ICT
     Neil Dixon and Nick Dixon 186

16   Language and talk in science education
     Richard Taylor 198

17   Literacy in science: a platform for learning
     Billy McClune 212

SECTION 4: Assessing science

18   A potted history of AfL in science
     Andrea Mapplebeck and Chris Harrison 226

19   Using data to inform science teaching
     Pete Robinson 235

20   International assessments of science: key points explained
     Yasmine El Masri 245

SECTION 5: Science teaching as a profession

21   Developing professional practice
     Alex Manning and Emma Towers 264

22   Mentoring new science teachers
     Michael Inglis 276

23   Leading science education in a secondary school
     science department
     Helen Gourlay and Euan Douglas 292

24   Health and safety in science
     Steve Jones 307

Conclusion 319

Index 320

FOREWORD

In the first 20 years of my career, I combined teaching and research as a university physicist. On one occasion, I presented a short talk to a conference describing my plans for an experiment that would explore a proposed extension of a well-known theory. In the ensuing discussion, one physicist commented that I was wasting my time, as established theory showed that the effect that I planned to look for could not possibly occur. Several months later, after investing time and resources in assembling the required equipment and in learning to handle the radioactive isotope involved, the first results seemed to demonstrate the predicted effect. But when the alignments of the equipment were altered to show an expected increase, the effect disappeared. There followed a brief and agonising re-think between my collaborator and myself – which concluded in triumph because, when we thought through the theory more carefully, we could see that the appearance in some conditions and disappearance in others were precisely the effects that we should have found.

This story can be summarised briefly as hypothesis, design and building an experiment, interpretation of first results, re-thinking of theory and of methods through discussion with colleagues, and confirmation of hypothesis – albeit in a refined form. All of these features can be experienced in the work of students in the secondary science classroom: they may all arise in a single laboratory inquiry, or the separate features may be explored in more restricted problem-solving tasks.

However, in classrooms where students are simply taught the facts and theories established by science, they will not encounter any of them, so they will have very little idea of the way that science increases our understanding of the natural world. Such limitations both miss the opportunities that science education can contribute to many aspects of the learning development of students, and will also mean that the students will not be as capable as they could be in making those judgements that all citizens may have to make about the work and the findings of scientists

There is a more specific and serious disadvantage. In making their choices of further study where science will be one of the options, students will be guided by their school experience. So that experience should give them an authentic impression of what it might be like to work as a scientist. Some university teachers have published very sad accounts of newly enrolled undergraduates being very distressed on discovering that the science subject they had chosen involved laboratory work that was quite unlike the ready-made exercises in which they had excelled in school – they had been misled. Their A-level results were invalid in that the inferences they had based on them were unjustified. Such invalidity will also mislead all those, whether in schools, in families, or in further and higher education, whose judgements have been based on those examination results.

The need for school science to give students an authentic experience of science is reflected in most of the chapters in this book. It overlaps with and has implications for many of the decisions that classroom teachers have to make. So it is not surprising that 24 chapters were deemed necessary to cover all of the many responsibilities of classroom teachers. Indeed, many teachers might find the long contents list quite daunting. However, the list itself does show that there are many types of interaction between the issues involved: one example is the interaction between the nature of school science and the motivation and engagement of students, another is the interaction between thinking about practical work and ensuring health and safety.

Teaching is itself a multi-dimensional learning experience, and one may continually change and shift between the core activities involved in helping students to engage in learning, to such secondary but essential activities as ensuring health and safety in the handling of apparatus. Across its 24 chapters, this book has been designed to help with all the dimensions, ranging from those that focus on the core aims of classroom teaching to those which, albeit of secondary importance, occasionally demand attention.

Teaching is the servant of learning, but is also itself a lifelong learning journey. This book should serve as a helpful companion throughout that journey.

Paul Black
Professor Emeritus of Science Education
King's College London
ASE Lifetime Achievement Award for Contributions to Science Education 2005

ABOUT THE AUTHORS

Ian Abrahams is Head of the School of Education and Professor of Science Education at the University of Lincoln. Having been a physics teacher, he obtained his PhD under the supervision of Robin Millar and, in 2007, moved into Higher Education. His research interests relate to practical work, misconceptions, academically asymmetrical peer mentoring, and the evaluation of educational interventions. Ian has directed the evaluation of various large scale national interventions, including the *Getting Practical* programme (2009-11), a Nuffield-funded project to improve the effectiveness of practical work (2012), and the Wellcome Trust-funded pilot CPD programme for primary science specialists (2012-14).

Indira Banner is a lecturer in science education at the University of Leeds. She taught science and biology in a secondary school in Bradford before joining her current university as research fellow in 2008, looking at the impact of the then new science curriculum introduced at Key Stage 4. She taught and led the Biology PGCE for many years and thoroughly enjoyed her visits to local schools. Now she is the Head of the Undergraduate Area and teaches on the BA Education and MA for practicing teachers and supervises doctoral students. Her research interests include students' attitudes towards, and participation in, science education. Indira is a member of the ASE Research Specialist Group.

Derek Bell is a teacher, researcher, adviser and advocate for improving and enriching science education for all. A past Chair of ASE, he worked in schools and universities before becoming Chief Executive of ASE and then Head of Education at the Wellcome Trust. He has written a wide range of publications and remains active in education, nationally and internationally, through his committee and advisory work.

Richard Brock is a postdoctoral fellow in science education at King's College London. Prior to this, he taught physics at a secondary school in England and has recently completed his PhD at the University of Cambridge, which focused on how students develop coherent conceptual structures related to concepts in physics. He has published articles on the role of insight and intuition in learning about science and on the use of the microgenetic method as a research approach in science education.

Helen Darlington CSciTeach is an innovative and experienced teacher of biology and psychology. In addition to her role as Head of Biology in her school, she contributes to conferences and has produced a range of publications. Her interest in educational neuroscience led to the completion of her PhD thesis in 2017, at UCL IoE, which explored the role of 'interest in science' among students at Key Stage 4 (age 14-16).

Euan Douglas CSciTeach is Head of Science at Saint George Catholic College in Southampton. He has worked in a range of secondary schools across Hampshire for over 10 years. He completed the Teaching Leaders Fellows Programme in 2015 and completed his NPQML qualification as part of this. He is a Chartered Science Teacher (CSciTeach) and has presented at ASE conferences within his region and nationally. He is a member of the ASE 11-19 Committee, as well as his regional committee where he has helped set up and run TeachMeets.

Lucy Dix taught in mainstream secondary schools for nine years, before completing her PhD considering the learning of children with Down's Syndrome in 2016. Since then, she has been teaching and researching in the field of inclusive education, disability studies and supporting children with special educational needs. She is now enjoying her return to the classroom.

Neil Dixon is a Specialist Leader of Education in Science, and the Director of Research, at South Bromsgrove High School in Worcestershire. This is a large non-selective academy and Teaching School, with a 13-18 age range. In addition to his teaching, Neil works as a subject expert for Ofqual, and has written textbooks and revision guides for science, as well as acting as a consultant and author for the BBC.

Nick Dixon is Head of Science at Magdalen College School in Brackley. This is a large, semi-rural, mixed non-selective academy in Northamptonshire. Nick has written revision guides and textbooks for science. He also works for the BBC as a consultant and author. With Neil, he leads an annual session for PGCE students at the University of Oxford on the use of ICT within science. Nick is a member of the Curriculum Committee of the Royal Society of Biology.

Yasmine El Masri is a Hulme Research Fellow in Educational Assessment at Oxford University Centre for Educational Assessment (OUCEA) and Brasenose College. She is currently leading an ESRC GCRF-funded project entitled *Using Technology in Science Tasks: Reducing Language Barriers for Syrian Refugees in Lebanon*. Her doctoral research investigated the impact of translation on item difficulty in PISA across three countries. Yasmine also teaches postgraduate modules on International Large Scale Assessment offered by the Department of Education at the University of Oxford. Before becoming a researcher, Yasmine was a science teacher in primary and secondary schools in Lebanon and the United Arab Emirates.

Nikolaos Fotou is a physicist and now a lecturer in science education at Maynooth University in Ireland. He has studied physics at the University of Crete in Greece and obtained his MA and PhD degrees in science education at the University of Leeds. In addition to doing research in the area of students' knowledge, reasoning and self-generation of analogies and the use of ICT in science and mathematics classrooms, he is interested in the role and effectiveness of practical work in secondary and tertiary science education.

Nicoleta Gaciu is a Senior Lecturer in Education at Oxford Brookes University. She has extensive experience in physics teaching and research, both at school and university level. Nicoleta has a diverse research background and interests that span education, statistics, physics, computing (particularly specialised in research methods to collect and analyse educational research data), promoting creativity and creative teaching and learning in science classrooms. She is a Convenor of the BERA Science Education Special Interest Group.

Melissa Glackin is lecturer in science education at King's College London. Melissa's research and teaching interests include teaching and learning science outside the classroom, teachers' beliefs and self-efficacy, in-service and pre-service teacher professional development and outdoor science and environmental education curriculum development. She teaches on the secondary Post Graduate Certification in Education programme (PGCE), where all science trainees experience a residential field course, and is subject director for the biology secondary science programme. Melissa has worked as a secondary school science teacher and on a range of outdoor learning projects.

Helen Gourlay is a Lecturer in STEM Education at Brunel University London, specialising in secondary physics education. She also leads the secondary PGCE programme. Before working in university partnership initial teacher education, Helen was a successful and innovative science teacher. She taught science across the secondary age range for 17 years in comprehensive schools in London and Hertfordshire. This experience included working as a Team Leader for Science in two schools. Prior to working at Brunel University, Helen was a lecturer at King's College London and at the University of East Anglia, where her responsibilities included leading the physics education aspect of initial teacher education, School Direct, and Teach First. Her research interests include science teacher development and children and young people's learning in science.

Mark Hardman is Lecturer in Science Education at the UCL Institute of Education. He has been a teacher educator and researcher for more than 10 years, prior to which he taught science in London. Mark has led a broad range of teacher education programmes across London and the South East, supporting undergraduate, postgraduate and postdoctoral student teachers as well as experienced science teachers. His research centres on how people learn within the messiness of classrooms, and he uses complexity theory, neuroscience and philosophical perspectives to better understand this.

Wynne Harlen OBE, PhD has been a teacher, teacher educator, researcher, curriculum developer and evaluator during her long career. She was Professor of Science Education and Head of the Education Department at the University of Liverpool, before moving to Scotland to become Director of the Scottish Council for Research in Education, 1990-99. She was Editor of *Primary Science Review* from 1999-2004 and President of the ASE in 2009. She has worked on several international projects and was Chair of the science expert group for the OECD's PISA project during its first six years.

Chris Harrison is Reader in Science Education at King's College London and a former Chair of ASE. She works closely with teachers promoting professional learning as a teacher educator and researcher. Her research in classroom assessment and inquiry learning has fed into much of what we now do in science classrooms, affecting both policy and practice. Chris has published widely for teachers and researchers and produced MOOCs to extend the impact of her team's research. She is passionate about bridging the research/practice divide and recognising and valuing practice-based evidence as well as evidence-based practice.

Judith Hillier is an Associate Professor in Science Education (Physics) at Oxford University Department of Education. After completing her PhD in condensed matter physics from the University of Leeds and the Institut Laue Langevin, Grenoble, she studied on the Oxford PGCE Internship scheme and then taught for several years in an Oxfordshire comprehensive school, becoming Key Stage 3 Co-ordinator. At Oxford, she leads the PGCE Science course, teaches on the Masters in Learning and Teaching part-time course for teachers, and runs the Teaching Physics in Schools option for 2nd year physics undergraduates. She is a founding member of the Oxfordshire Schools Physics Partnership, sponsored by the Ogden Trust, and supervises a number of PhD students. Her research interests include factors influencing the recruitment and retention of physics teachers in the teaching profession, knowledge and professional practice of pre-service physics teachers, explanations in science education, pedagogical practices in physics education and supporting the learning of disadvantaged students in science.

Michael Inglis is a Lecturer in Science Education and member of the Centre for Studies in Science and Mathematics Education at the University of Leeds. He has been a science PGCE tutor for 10 years and leads the BA Education programme. His research is in teachers' experiences of initial teacher education and physics subject knowledge development. Michael taught physics in secondary schools in London and Hertfordshire, and supported the professional development of science teachers and teacher educators in Ethiopia for two years with Voluntary Service Overseas (VSO). Prior to becoming a teacher, Michael was a geophysicist in the oil industry.

Steve Jones taught science in secondary schools for 18 years before moving into advisory work, initially with Hertfordshire and then as a regional adviser with the Secondary National Strategies. Always passionate about practical work, Steve believes that hands-on activities are quite simply the best way for students to learn science. Moving to CLEAPSS, where he is now Director, in 2011 enabled him to indulge his interest further and to develop a strong commitment to proportionate and common sense approaches to managing risks in practical subjects

Billy McClune is senior lecturer in the School of Social Sciences, Education and Social Work at Queen's University Belfast, where he co-ordinates the PGCE science programme. He taught physics for many years in Northern Ireland and was the Head of Science Department before taking a role in initial teacher education. His interest in science literacy with a particular focus on science in the news media has resulted in a number of school-based research studies in Northern Ireland. These studies have been informed and complemented by periods of study and school-based observation in the United States, Canada and New Zealand.

Deb McGregor is a Professor at Oxford Brookes University. She has been a classroom teacher, advisory teacher, teacher educator and educational researcher. She has led a variety of science education projects in both primary and secondary classrooms. She has been a member (and Chair) of the ASE Research Specialist Group. She is passionate about ways in which school science should be engaging and participatory so that thinking and creativity become part of the natural learning process.

Alex Manning Before joining King's College London, Alex worked in London schools as a secondary science teacher and Head of Year. She joined King's in 2003, and has recently completed her PhD. While her teaching responsibilities are primarily concerned with the recruitment of teachers and initial education, her research focuses on the bigger picture of teacher retention. She is interested in teachers' motivations for joining the profession, how these relate to notions of resilience and how teachers can be supported to remain in the classroom.

Andrea Mapplebeck CSciTeach worked in a variety of roles in secondary schools, and has been writing and facilitating hands-on and evidenced-based CPD for schools, teachers and other organisations across the UK and abroad since 2002. Andrea has co-produced a number of MOOCs focusing on aspects of formative practice and supported other online learning CPD activities. Andrea's research has involved evaluating the impact of professional development on teacher's practice and she is currently studying part-time for a PhD, looking at the characteristics of oral teacher feedback that move forward student learning in a science classroom.

Brian Matthews taught in London schools before leading the PGCE at Goldsmith's College. At King's, he contributes to the science PGCE programme and was involved in research funded by the EU into *Strategies for Assessment of Inquiry Learning in Science* (SAILS). Previously he led a research project on emotional engagement in science education. Brian's research interests include science inquiry, gender and young people's emotions (www.engaging education.co.uk). His publications include *Engaging Education: Developing emotional literacy, equity and co-education* (2016), McGraw-Hill/OUP.

Richard Needham CSciTeach has wide and extensive experience of science education, beginning his career as a biology teacher and then senior leader in a large secondary school. More recently he has held positions as a senior lecturer in science education, a business development manager with New Media educational software and was educational technology lead at the National Science Learning Centre. Richard is now an education consultant specialising in curriculum and professional development for all engaged in science education. He is Chair of Trustees for ASE and a member of ASE's Registration Board for Chartered Science Teachers.

Pete Robinson CSciTeach has a national profile within the UK for his work with ASE and as the Chair of the Association from 2013-2014. Pete has worked as an independent teaching and learning consultant since 2010, fulfilling a variety of roles within the UK and abroad. Prior to that, Pete was the National Strategies consultant for Bury Local Authority for eight years. Between 1983 and 2002, Pete worked as a classroom teacher, Head of Physics and Head of Science in three schools in different parts of England.

Mike Sands is a science teacher with a chemistry specialism who has worked in a variety of schools over the last 15 years, and who has experience in a variety of posts within science departments. He is currently working in an inner city Hull academy as a Lead Teacher of Chemistry. He has experience of teaching all key stages and is an experienced examiner, having worked with several exam boards from Key Stage 3 to Key Stage 5 (age 11-19). He was awarded National STEM Expert Teacher in 2015 by STEM Learning at the National STEM Centre in York.

Natasha Serret was a primary science teacher in London for six years, before joining King's College London where she worked as a senior researcher for 10 years, working on the primary CASE project, *Thinking Beyond the Classroom* and KOSAP (King's-Oxfordshire-Summative-Assessment Project). She is currently a senior lecturer at Nottingham Trent University and leads on science for all teacher education routes. She recently worked as a postdoctoral researcher at King's on EU-FP7 ASSISTME (Assess Inquiry in Science Technology and Mathematics Education).

Keith Taber CSciTeach is the professor of science education at the University of Cambridge. Keith taught chemistry and physics in secondary schools and further education, and has worked supporting graduates entering science teaching, responsible for the PGCE physics course at Cambridge. He currently teaches educational research and supervises research students. He is the Editor of the free-to-access journal *Chemistry Education Research and Practice*, and edited the ASE's practice handbook on *Teaching Secondary Chemistry* (2012). He has published widely, including *Student Thinking and Learning in Science: Perspectives on the nature and development of learners' ideas* (Routledge, 2014).

Richard Taylor is a physics teacher at Gillingham School in Dorset, with a physics degree from the University of Southampton and a PGCE from the University of Exeter. He has taught science in Key Stages 3, 4 and 5 for over 10 years, has led Key Stage 3 Science and provided professional development for teachers at various stages of their careers. He has a Masters' degree in Teaching and Learning from the University of Oxford, with his research being focused on the role of language in learning science. He is currently extending this research as he studies for a PhD in Education, again at the University of Oxford.

Clare Thomson taught physics in a variety of schools, both single sex and co-ed, in London and the North West, ending up as a Head of Department. During that time, she completed a MEd in science education at King's College London. More recently, she worked at the Institute of Physics, managing a number of projects exploring diversity and gender equity in school physics. She continues to have an interest in work in this area. She is the recipient of an ASE Special Service award for her work with the London Region committee and is on the editorial board of the international journal, *Physics Education*.

Emma Towers taught for 10 years in a London primary school before joining King's College London in 2012 to complete her PhD. Her doctoral study explored the reasons why teachers and Headteachers stay working in inner London primary schools. At King's, she teaches on undergraduate and postgraduate courses. Her main research interests include teachers' lives and career trajectories, urban education and teacher identity.

Ed Walsh is a curriculum developer, CPD provider and teacher trainer. A science teacher for 20 years and a team leader for 12 of those, he was Science Adviser for Cornwall and Senior Adviser for the National Strategies. He is Series Editor for Collins KS3 and GCSE materials, CPD provider for AQA, Regional Development Lead for the National STEM Learning Centre and Senior Regional Hub Leader for Primary Science Quality Mark. Ed tutors on primary and secondary SCITT programmes and runs workshops at regional and national ASE events.

Helen Wilson is a Principal Lecturer in Science Education at Oxford Brookes University. She focuses her research into the links between creative, challenging primary science lessons and the resulting increase in pupils' attitudes and attainment. She jointly leads the *Thinking, Doing, Talking Science* project, funded by the Education Endowment Foundation.

Fiona Woodhouse A lifelong desire to teach saw Fiona completing her PGCE after her first degree. She spent 15 years in schools in West Yorkshire teaching mainly biology and science. She then moved to the School of Education at the University of Huddersfield, where she has been both the PGCE Course Leader and Science Lead Tutor, inspiring a new generation of science teachers. She also has been able to follow the careers of many of these teachers, supporting them with Masters' degrees and PhDs. She feels that it has been a delight and privilege to work with so many outstanding science teachers for over 30 years and ASE has been a constant support during this time.

ABOUT THE ASSOCIATION FOR SCIENCE EDUCATION (ASE)

A professional network for science educators

The Association for Science Education (ASE) is a professional community that supports teachers and technicians to develop excellent science teaching and learning. Join our network today!

11-19 membership benefits include:

- ➤ Professional development: The ASE community puts together events from TeachMeets to workshops, to our flagship Annual Conference, the largest subject teaching event in Europe. ASE members receive substantial discounts, often as much as 50% off the delegate rate.
- ➤ School Science Review and Education in Science: ASE's journals offer invaluable insights into educational research, practical ideas, news and information, and examples of good practice. Recent issues have covered the new GCSEs, the use of mathematics in science, the impact of curriculum change and space science. Members can access the full archive online.
- ➤ Public liability insurance: ASE membership covers you in the classroom or prep room, for practical and school trips up to £5m.
- ➤ And lots more! Members have the opportunity to get involved with local events, national committees or sharing views and ideas. There are substantial discounts for members in the ASE Bookshop, which offers a range of highly recommended titles for science educators and all those interested in school science education.
- ➤ Trainee teacher special offer: Receive one free CPD day at the ASE Annual Conference and purchase this Guide at a special price!

Explore our resources at www.ase.org.uk/resources. To join ASE, please visit our website at www.ase.org.uk

The **Association** for **Science Education**
Promoting Excellence in Science Teaching and Learning

INTRODUCTION

We are delighted to present this updated edition of the *ASE Guide to Secondary Science Education,* which has been written for all those involved in secondary school science education. It is intended to be a handbook to support science teachers and those working with them in and out of school, both at the start of their teaching career and in years to come. All the authors are writing from a UK perspective, with examples often being drawn from classrooms in England. However, the good practice described here is valid and applicable to the majority of contexts across the UK, and indeed the world, and therefore should be useful to science educators from a broad set of backgrounds.

This book covers a wide range of topics, from considering why and how we might teach science, to engaging and motivating the diverse student community with whom we work, to how to mentor new colleagues. Some of the chapters offer an engaging overview of important theoretical ideas, which form a key part of our professional knowledge as teachers. Other chapters are more focused on developing strategies to use in the classroom. All the chapters offer opportunities for teachers at various stages in their careers to question and reflect on their practice in order to continue their professional learning, including suggested further reading as and when the reader might want to pursue ideas in greater depth.

Each author has been careful to introduce and explain key terms and acronyms, but readers should bear in mind the following aspects of the education system in England: primary school is for students aged 5-11 years, with secondary school for students aged 11-18 years. These are then divided further into key stages, with Key Stage 1 for students aged 5-7 years, Key Stage 2 for ages 7-11, Key Stage 3 for ages 11-14, Key Stage 4 for ages 14-16 and Key Stage 5 for ages 16-18.

Below is an overview of the organisation of this book to help guide the reader, although some of these groupings are somewhat artificial: for example, every chapter has learners and the principle of inclusion at its heart. The book can be read from cover to cover, but equally each chapter can stand alone, with the reader choosing as and when to engage with the various ideas.

Section 1: Foundations of science education

As science teachers, it can be important to stand back and consider why we teach science and the nature of the subject that we are teaching. It is also essential to think about what we know about the process of learning and our students' previous experiences of science education. These foundations enable each teacher to develop his/her own philosophy of and approach to teaching science.

Section 2: Students: all learning science

The students who we teach are all individuals who come with a range of attitudes towards and aptitudes in science, language and number, and different levels of confidence towards school and their own learning. It is important to develop a deeper understanding of our students as learners, their starting points and interests, and differences between them. The subsequent implications for our professional practice are crucial for effective inclusive teaching and learning of science.

Section 3: Science teachers: synthesising learning

Science is a multi-faceted discipline, and teaching science in school is no less complicated. Science teachers' professional knowledge enables them to consider the literacy and mathematical demands of the curriculum, the range of skills for learners to develop, and the opportunities afforded by a wide range of teaching and learning strategies. This section provides support in reflecting on these aspects of teaching, so all this expertise is brought together when planning and teaching science lessons.

Section 4: Assessing science

Assessment is an important part of the teaching and learning process, allowing teachers and learners to evaluate the effectiveness of their endeavours and to identify the next steps. The challenge is to accomplish this in a manageable and meaningful way, whilst also recognising the contributions and limitations of collecting and using data and the international discussions about assessment.

Section 5: Science teaching as a profession

The professional nature of teaching is both well recognised and widely debated, encompassing both clearly defined skills such as meeting health and safety requirements, and more nebulous, but equally important, aspects including working well with colleagues, supporting them and also maintaining one's own work-life balance and wellbeing. Considering what it means to become part of the science teaching profession is an important part of one's integration into this mainstay of our education system.

The *ASE Guide to Secondary Science Education* is one of a series of books, with the other publications being the *ASE Guide to Primary Science Education* and the *ASE Guide to Research in Science Education*, with new editions of these titles accompanying or following the publication of this book.

Acknowledgements

We are extremely grateful to all the authors who so kindly agreed to contribute to this book: for their careful following of the briefs we gave, their gracious responses to our feedback and prompt replies to all our queries and requests for further information.

Thank you to Jane Hanrott for her hard work turning the collection of files into the final publishable book. Thank you to Marianne Cutler, Alaric Thompson and the ASE Publications Specialist Group for their feedback and guidance.

Finally, thank you to Chris Harrison, both for her help with the final editing, and for asking us to do this in the first place. It has been a highly interesting and mostly very enjoyable task, and hopefully the fruits of our labours will be worthwhile and of benefit to all those who read this *Guide*. We have endeavoured to ensure that this book is accurate throughout and apologise for any remaining errors, for which we accept responsibility.

Indira Banner and **Judith Hillier**
University of Leeds and University of Oxford
July 2018

SECTION 1:

Foundations of science education

CHAPTER 1

Aims and approaches of good science education

Wynne Harlen

At a time when developments such as the use of ICT and social media are changing what happens in the classroom as much as life outside, it is important to reflect on and affirm what we are trying to achieve. This chapter considers the enduring aims of science education and the arguments for their importance. Implications for practice arising from the nature and range of these aims are discussed in sections on pedagogy, assessment and curriculum content. Most attention is given to pedagogy (teaching) and particularly to approaches that are based on current views of how learning takes place: constructivism, discussion and dialogue, and inquiry. The combination of these approaches identifies good teaching, which, together with effective use of assessment and curriculum content aimed at developing big ideas, we might say defines 'good science education'.

Introduction

Whether we are designing activities for students, a programme of work for a year, or a national curriculum, it is good practice to start by establishing the aims and purposes to be achieved. So, thinking of science education as a whole, what are we aiming to achieve? There is no lack of statements of the importance of science education and the learning it should aim to achieve (e.g. AAAS, 1989; Fensham, 1985; Millar & Osborne, 1998; NRC, 2012; OECD, 2016). Looking across these sources, they have much in common in aiming to develop in students:

- ➤ understanding of 'big', or key, ideas, facts and principles (sometimes called ideas *of* science or content knowledge);
- ➤ understanding of the nature of science (sometimes called ideas *about* science or epistemic knowledge); and
- ➤ capability in the practices of science concerned with gathering and using evidence (sometimes known as inquiry skills or procedural knowledge).

Some sources, in particular OECD (2016), refer to 'scientific literacy', which encompasses these three aims in a formal definition used in the PISA science surveys. Expressed more informally, scientific literacy places emphasis not on mastering a body of knowledge, but on having – and being able to use – a general understanding of key ideas of science, and about the nature and limitations of science that can be used in making decisions and participating in society as an informed and concerned citizen.

Harlen (2015: 7) also adds the aim to *develop and sustain learners' curiosity about the world, enjoyment of scientific activity'*. This affective dimension is important in influencing motivation to sustain engagement in scientific activity and to take interest in scientific topics. In addition, of course, there are general aims that apply to all areas of the curriculum. These include the ability to continue learning, in recognition that today's students will have to make more choices than those living in past decades. They will undertake tasks and have occupations that we cannot foresee. They will need to *'become able to organise and regulate their own learning, to learn independently and in groups and to overcome difficulties in the learning process'* (OECD, 1999 p.9).

Rationale – why these aims?

Although their value may be self-evident to many readers, it is worth briefly reviewing reasons for selecting these aims. A grasp of the key ideas has benefits for students as individuals and for society:

➢ The benefits for individuals of developing powerful ideas that have wide application follow from being able to grasp the essential features of events or phenomena, even though lacking knowledge of every detail. Understanding aspects of the world around helps individuals in their personal decisions that affect their health and enjoyment of the environment, as well as their choice of career. The practice of questioning, seeking evidence and answers, and sharing views with others also contributes to building confidence and respect for themselves and others. Furthermore, the satisfaction of being able to see patterns in different situations and connections between them provides important motivation for learning during and beyond formal education.

➢ Benefits for society follow from young people being able to make informed choices both as students and later in life about, for instance, their diet, exercise, use of energy and care of the environment. As well as impact on their own daily lives, such matters have wider implications for their own and others' future lives through the longer-term impact of human activity on the environment. Understanding how science is used in many aspects of life is needed for appreciating the importance of science and for recognising the attention that needs to be given to ensuring that scientific knowledge is used appropriately.

Knowledge of the nature of science enables students to know how the ideas that explain things in the world around have been arrived at, not just what these ideas are. Without knowing how ideas were developed, learning science would require blind acceptance of many ideas about the natural world that appear to run counter to common sense. In a world increasingly dependent on the applications of science, people may feel powerless without some understanding of how to evaluate the quality of the information on which explanations are based.

Developing capability through engagement in scientific inquiry also helps understanding of how ideas are developed through scientific activity. The key characteristic of such activity is an attempt to answer a question to which students don't know the answer, or to explain something that they don't understand. The answer to some questions can be found by first-hand investigation but, for others, information is needed from secondary sources. In either case, the important feature is that evidence is used to test ideas, as part of a process that is best understood through participation in it. So, developing capabilities involved in conducting scientific inquiry has a key role in the development of ideas both *of* science and *about* science.

Implications of these aims

Moving from rationale to implementation, if we take these aims seriously there are implications for all aspects of students' experience in school science: for the curriculum content; for pedagogy; and for assessment. These aspects are not independent of one another; changes in one affect the others. It is no use suggesting that the content should be focused on big ideas if the assessment requires memorising multiple facts, or if the pedagogy does not forge links that are necessary to form these big ideas. It is no use advocating the employment of inquiry-based teaching if there is an overbearing summative assessment system, or a curriculum overcrowded with content. Nor can we expect students to develop responsibility for their own continued learning if teaching does not allow time for reflection and room for creativity, nor hope for positive attitudes towards science if the curriculum content seems, to students, to be remote from their interests and experience.

Although the intention here is to focus on teaching approaches and tasks (pedagogy) – since this is the aspect that is most in the control of practitioners – with these interactions in mind, we will not neglect implications for assessment and curriculum content.

Pedagogy – how we teach

All three of the overall aims stated earlier force consideration of pedagogy. Innovative approaches to teaching are regularly developed and advocated, some producing enduring change in teaching, others disappearing after the initial novelty has worn off. Those that are firmly based on understanding of how learning takes place have more chance of becoming embedded in practice. They include the following:

➤ individual and social constructivism;

➤ dialogue, discussion and argumentation; and

➤ inquiry.

These are in no way rivals or in competition with each other; indeed, they overlap considerably and would benefit from being brought together to create a more coherent – and scientific – approach to teaching and learning. In combination, they could constitute a strong pedagogy for achieving scientific literacy and the skills needed for lifelong learning. We look at each separately before trying to envisage what a combination might be.

Individual and social constructivism

Constructivism derives from the well-established evidence (Driver, 1983; Osborne & Freyberg, 1985; SPACE, 1990–98) that children work things out for themselves from an early age and often arrive at ideas that conflict with scientific ones, because they are based on the children's necessarily limited experience and reasoning. But, seen from the children's point of view, they are reasonable. These ideas are not to be stamped out by telling the students the 'right' answer, which would in most cases be too complex for them to understand.

Constructivist pedagogy starts from these existing ideas and sees the role of the teacher as providing students with the experiences, evidence and reasoning skills that will lead them to more scientific ideas. However, a constructivist approach has implications beyond knowing students' existing ideas. It identifies effective learning as involving active participation of the learner, distinguishing it from a view of learning as the acquisition of more knowledge and skills. In the important matter of helping learners to consider alternative ideas more consistent with the scientific view, there are various strategies open to teachers, such as: extending experience; helping students to test ideas; linking ideas from one experience to a related one; and introducing alternative ideas.

More recently, we have realised that learning is not entirely, or even mainly, an individual matter, but takes place through social interaction. This is the basis of the social or *sociocultural constructivist* perspective of learning. In this view, understanding results from making sense of new experiences with others, rather than by working individually. In a collaborative group, an individual learner takes from a shared experience what is needed to help his or her understanding, then communicates the result as an input into the group discussion. There is a constant to-ing and fro-ing from individual to group, as knowledge is constructed together through social interaction and dialogue. Physical resources and language also have important roles in this process (James, 2012; Harlen, 2013 p.32).

The view of learning as a social and collaborative activity recognises the value of talk. This makes a link to the next approach to be considered. But first it is useful to note that, although in classrooms the interaction among students and between students and teacher is mostly face-to-face, learning from and with others can be through the written word as well. Feedback to students in writing (marking) can be an effective channel for dialogue between teacher and students, as long as the comments take learning forward and that students have time to read and reflect on them. However, talk is an important vehicle for learning and, in various forms, is central to teaching approaches, as we now see.

Dialogue, discussion and argumentation

It is through language that we develop a shared understanding of ideas. Communicating our ideas involves trying to find words that convey our meaning to others. In this process, our own ideas often have to be reformulated in ways that are influenced by the meaning that others give to words. So, the very act of talking can change our understanding. The value for learning of talking was established in ancient times, but, in modern times, it was Douglas Barnes in the 1970s whose research led to the realisation of the importance of informal or 'exploratory' talk. In this kind of talk, later elaborated upon by Asoko and Scott (2006) and by Mercer *et al* (2004), students interrupt each other, repeat themselves, hesitate and rephrase. Barnes suggested that students only engage in this kind of talk in the absence of the teacher, because then there is no source of authority to which students can turn. However, Alexander (2008) has identified a role for the teacher in such verbal interactions, in the form of 'dialogic teaching', described as *a distinct pedagogical approach, which harnesses the power of talk to stimulate and extend children's thinking, and to advance their learning and understanding. It also enables the teacher more precisely to diagnose and assess'* (Alexander, 2008: 1). The aim of dialogic talk, of exploring a subject in depth, distinguishes it from the chatter of the playground.

In relation to science, the teacher's role in dialogic teaching is to encourage students to explain their thinking, to focus on the use of evidence and to engage in what has been described as 'argumentation'. This is different from argument in daily life:

'In science, goals of argumentation are to promote as much understanding of a situation as possible and to persuade colleagues of the validity of a specific idea. Rather than trying to win an argument, as people often do in non-science contexts, scientific argumentation is ideally about sharing, processing and learning about ideas' (Michaels et al, 2008 p.89).

Inquiry

Inquiry, like dialogue, has a meaning in the context of science education that is different from its everyday usage. It refers to a process in which students develop their understanding through collecting and using evidence to test ideas. The process begins with a question to be answered, a problem to be solved or some new experience to be explained. Faced with trying to make sense of the new experience or problem, students do what we all do in such a situation, that is, to try to use ideas from previous experience. Initial exploration may identify features that bring to mind ideas that may provide a solution (*'I think it might be…'*, *'I've seen something like this when…'*, *'It's a bit like…'*). There might be several ideas that could be relevant and, through discussion, one of these is chosen as a hypothesis to be tried, to see if there is evidence supporting it. Testing the hypothesis scientifically will involve making predictions, gathering, analysing and interpreting new data and communicating results. The source of new data may be the direct investigation of objects or materials, observation of events or use of secondary sources accessed through the Internet or books. If the new evidence does not support the prediction, then an alternative idea needs to be tried. When new evidence does support the initial idea, then that idea becomes strengthened because it then explains the new phenomenon as well as earlier experience. In this way, ideas change from being 'small' (just explaining a certain event) to being 'bigger', since they explain a greater number of events.

Understanding, built through this activity, depends crucially on the inquiry skills – on how predictions are made, what evidence is collected and how it is interpreted, hence the importance of developing capability in using science inquiry skills. Interpretation of data to provide evidence to test ideas may involve debate with other students and the teacher and finding out what experts have concluded. Implicit in all of this is that students are taking part in activities similar to those in which scientists engage in developing understanding. By making these activities conscious, students develop their ideas about the nature of scientific activity.

A renewed pedagogy

The three approaches to teaching that we have considered complement each other by each giving emphasis to aspects that receive less emphasis in the other approaches:

> Constructivism lays emphasis on starting from the ideas that students bring from their previous experience.
> Dialogue stresses the value of talking, listening and sharing in the development of ideas.
> Inquiry identifies the importance of teaching ideas by collecting and using evidence.

In combination, they go some way to defining effective pedagogy, but this is only part of what is needed for working effectively towards the aims set out at the start of this chapter. As we noted earlier, this also depends on two further aspects of students' learning experiences: assessment and curriculum content. Here we can give only a brief comment on these topics. The suggestions for reading at the end of this chapter take this further.

Formative and summative assessment

'Formative' and 'summative' identify two of the main purposes of assessment:

> to help learning, and
> to report on what has been learned at a particular time.

We consider briefly here how the use of assessment for these purposes influences the achievement of the aims of science education.

Formative assessment (or assessment *for* learning) is an ongoing part of everyday practice and could well be considered as an aspect of pedagogy. It shares with inquiry the aim of developing understanding through learners taking charge of their learning. The formative use of assessment is a continuing cyclic process in which information about students' ideas and skills informs ongoing teaching and helps learners' active engagement in learning (Harlen & Qualter, 2014). It involves the collection of evidence about learning as it takes place, the interpretation of that evidence in terms of progress towards the goals of the work, the identification of appropriate next steps and decisions about how to take them. It helps to ensure that there is progression and regulates the teaching and learning processes to ensure learning with understanding, through feedback to both teacher and student. Formative assessment is also central to enabling students to acquire ownership of their learning, one of the key features of genuine understanding. The role of teachers in using assessment in this way is not only to find out where students are in relation to the goals, and to provide activities with the right amount of challenge to advance their existing ideas and skills, but also to share the goals with students and help them assess their own progress towards them.

Formative assessment is focused on the short-term goals of lessons or sequence of lessons on a topic and has the role of helping the achievement of these goals. Summative assessment (or assessment *of* learning) focuses on medium and longer-term goals and achievement over time, not outcomes that can be achieved in a few lessons. What is assessed at the end of a term or year should reflect all the learning goals: not just easily tested facts and principles, but the big ideas to which these are linked, the developing grasp of the nature of science, and progress in using science inquiry skills. Summative assessment is important because what is assessed and reported signifies what is considered important to learn and is often taken as a guide to what is taught. Of course, we want to be sure that basic facts are secure, but this is easily tested. If other goals are to be taken seriously, then they should be included in what is assessed and reported.

Curriculum content

Decisions about curriculum content in science education face considerable challenges in light of the knowledge explosion. The daily addition to our knowledge about the living and made world – widely accessible through television programmes and other media reports about newly explored parts of planet Earth and indeed other planets – is one sign of the rapid increase in scientific knowledge. Other signs are in the applications of science in constantly changing technology, particularly in our modes of communication and access to information. These events raise important questions that must not be avoided:

> - How can science education be expected to keep up with this knowledge explosion?
> - Is it inevitable that what is taught in schools will be seen to be out-of-date and out of touch because events move more quickly than curricula and learning materials can be changed?
> - Isn't the attempt to 'cover' too much content bound to lead to short-term memorisation rather than deeper learning?

Part of the answer is to change the way we conceive and communicate goals of the curriculum. We are less at the mercy of constantly expanding information if we think of the aims of science education not in terms of a collection of facts and theories, but as a progression towards the development of broad underlying ideas that have wide and enduring application. This is the reason for the references in this chapter to 'big' ideas in discussing aims and goals.

These ideas, described as 'big' or powerful ideas, are ones that help understanding of a wide range of phenomena. 'Big' ideas are built from 'small' ideas that relate to specific objects or events. For example, the idea that earthworms have features that enable them to survive underground is a small idea. This idea gradually expands as it becomes linked to ideas from study of other organisms and is developed into a generalisation that applies to all organisms – a 'big' idea that endures no matter what new species are discovered.

In the context of school science it is important for teachers, when helping children to develop the small ideas, to see these as steps towards the 'bigger' ones. But how can the most important and powerful ideas be identified? For a start, there is no single 'right' list of big ideas to be uncovered. The selection of ideas is bound to depend on human judgement. The people in the best position to judge are those with experience and expertise in science and science education, and in related fields of technology and engineering where scientific ideas are used. Establishing criteria that such ideas would have to meet is an important starting point. As an example, criteria agreed by an international group of science teachers, scientists, meeting for the purpose of identifying key ideas, were that such ideas would:

> - have explanatory power in relation to a large number of objects, events and phenomena that are encountered by students in their lives during and after their school years;

> ➤ provide a basis for understanding issues, such as the use of energy, involved in making decisions that affect learners' own and others' health and wellbeing and the environment;

> ➤ lead to enjoyment and satisfaction in being able to answer or find answers to the kinds of questions that people ask about themselves and the natural world; and

> ➤ have cultural significance – for instance in affecting views of the human condition – reflecting achievements in the history of science, inspiration from the study of nature and the impacts of human activity on the environment (Harlen, 2015 p.14).

The outcome of applying these criteria was the identification of 10 big ideas of science and four ideas about science. These ideas and the progression towards them can be downloaded from the ASE website (see *Further reading*).

In conclusion

In this chapter, we have implicitly identified 'good science education' in terms of aims of learning in science and the means to achieve these aims. The means relate to pedagogy, assessment and curriculum content. Good teaching (pedagogy) combines approaches that enable students to start from their existing ideas and skills and build their understanding through discussion and inquiry. In addition, for good science education there needs to be assessment that is consistent with, and helps the achievement of, the aims of learning. There is also need for a curriculum expressed in terms of key ideas, which can be achieved through activities and topics that are seen as relevant and worthwhile by students and which help them to enjoy scientific activity.

Further reading

Osborne, J. & Ratcliffe, M. (2002) 'Developing effective methods of assessing ideas and evidence', *School Science Review*, **83**, (305), 113–123

Naylor, S. (2015) 'Talking and thinking using concept cartoons: what have we learned?', *School Science Review*, **97**, (359), 61–67

Harlen, W. (Ed.) 2015 *Working with Big Ideas in Science Education*. Trieste: IAP SEP. (Downloadable from www.ase.org.uk/documents/working-with-the-big-ideas-in-science-education/)

Harlen, W. (2013) *Assessment and Inquiry-Based Science Education*. Trieste: IAP Science Education Programme. Available from: www.interacademies.net/File.aspx?id=21245

References

AAAS (American Association for the Advancement of Science (1990) *Science for all Americans. Project 2061.* Oxford: Oxford University Press

Alexander, R. (2008) *Towards Dialogic Teaching: Rethinking Classroom Talk* (Fourth Edition). Cambridge: Dialogos

Asoko, H. & Scott, P. (2006) 'Talk in science classrooms'. In *ASE Guide to Primary Science Education,* Harlen, W. (Ed.) Hatfield: Association for Science Education (158–166)

Barnes, D. (1976) *From Communication to Curriculum.* Harmondsworth: Penguin

Bransford, J.D., Brown, A.L. & Cocking, R.R. (Eds.) (1999) *How People Learn: Brain, Mind, Experience and School.* Washington, DC: National Academy Press

Driver, R. (1983) *The Student as Scientist?* Milton Keynes: Open University Press

Duschl, R.A., Schweingruber, H.A. & Shouse, A.W. (Eds.) (2007) *Taking Science to School: Learning and Teaching Science in Grades K–8.* Board on Science Education Center for Education, Division of Behavioural and Social Sciences and Education. Washington DC: The National Academies Press

Fensham, P. (1985) 'Science for all: a reflective essay', *Journal of Curriculum Studies,* 17/4: 415-435

Harlen, W. & Qualter, A. (2014) *The Teaching of Science in Primary Schools.* London: Routledge

Harlen, W. (2008) 'Science as a key component of the primary curriculum: a rationale with policy implications', *Perspectives on Education 1 (Primary Science):* 4–18

Mercer, N., Dawes, L., Wegerif, R. & Sams, C. (2004) 'Reasoning as a scientist: ways of helping children to use language to learn science', *British Educational Research Journal,* **30,** (3), 359–377

Michaels, S., Shouse, A.W. & Schweingruber, H.A. (2008) *Ready, Set, Science! Putting Research to Work in K–8 Science Classrooms.* Washington: National Academies Press

Millar, R. & Osborne, J. (Eds.) (1998) *Beyond 2000. Science Education for the Future.* Available for download at: http://www.nuffieldfoundation.org/sites/default/files/Beyond%202000.pdf/

NRC (National Research Council) (2012) *A Framework for K-12 Science Education: Practices, Crosscutting Concepts and Core Ideas.* Washington DC: The National Academies Press

OECD (Organisation for Economic Co-operation and Development) (2016) *PISA 2015 Assessment and analytical Framework.* Paris: OECD

OECD (1999) *Measuring Student Knowledge and Skills: A New Framework for Assessment.* Paris: OECD

Osborne, R.J. & Freyberg, P. (1985) *Learning in Science: the Implications of 'Children's Science'.* Auckland: Heinemann

SPACE Reports (1990–1998) Titles include: *Evaporation and condensation* (1990), *Light* (1990), *Growth* (1990), *Electricity* (1991), *Materials* (1991), *Processes of Life* (1992), *Rocks, soil and weather* (1993), *Earth in space* (1996), *Forces* (1998). Liverpool: University of Liverpool Press

CHAPTER 2

The nature of science and the nature of school science

Keith S. Taber & Richard Brock

It has been widely argued that school science should incorporate more emphasis on teaching and learning about the nature of science. This chapter will briefly review the argument for such a position, consider what is meant by the nature of science, and highlight some of the challenges to incorporating this theme within the secondary school science curriculum.

The importance of teaching about the nature of science

There are several rationales for teaching about the nature of science. One logic of a curriculum organised by discipline-based school subjects (which is certainly not the only way a curriculum might be designed) is that education should introduce young people to the different forms of cultural activity in a society – and this would clearly include science. Each discipline has its own modes of thinking, of considering evidence, for example, and these offer important tools for adult life.

As children spend a decade or more in school being required to learn subjects labelled as 'science', 'mathematics', 'history', and so forth, we should surely expect them to come to appreciate the nature of the subjects that they have been studying for all this time. We might think that they could acquire this appreciation by a kind of cognitive osmosis, even when it is not made explicit. However, there is plenty of research to suggest that many students only acquire a vague notion of the nature of science, despite all those hours in science lessons. Even effective learners of specific school science topics may struggle to explain what makes one activity, but not another, science. This is a little like someone carefully learning the rules for how each chess piece may move on a chess board – without ever seeing or playing an actual game of chess. Similarly, in science lessons, students often miss the big picture.

It seems that school science often does not offer an authentic representation of actual scientific practice (we consider some of the challenges here later in the chapter). Yet, appreciating the nature of science as an area of professional activity is rather important for making informed choices relating to further study and possible careers. Scientists may need to be creative and insightful, effective problem-solvers, tenacious, potentially single-minded (when needed) and potentially open-minded (also, when needed), patient, pragmatic, able to follow procedures carefully yet also to have the confidence to sometimes disregard them. These qualities may not all be emphasised (or even encouraged) by school science courses.

School science is not just about preparation for those who want a career in science. A key function of school science is to prepare all young people for life in technologically advanced societies that face major challenges. A good science education should provide people with the ability to understand and evaluate evidence and arguments that they meet in their everyday lives – and support their decision-making in areas such as consumption (what to spend their money on), health (when alternative treatments and likely prognoses are presented to them), lifestyle (such as behaviours relating to recycling and pollution) and civil engagement (when science is offered to support arguments for and against policies offered by politicians). This requires a science education that includes core scientific ideas and an appreciation of the key processes of science.

Learners need to appreciate how science offers robust evidence-based guidance based on uncertain knowledge. Scientific knowledge is always (technically) provisional – open to being reconsidered in the light of new evidence. This appreciation is vital when people are asked to engage in political action (e.g. voting) on issues where scientists may appear to publicly disagree, such as the causes, extent and rate of global warning, the dangers of extensive deforestation and the relative risks and benefits of nuclear power generation.

Science and school science

A common aspect of school science is that it tends to focus on the areas of knowledge that are currently considered stable, whereas most research activity in science takes place in areas where there is no strong consensus yet, because more research is needed to better understand the phenomena – what has been termed science-in-the-making or science in action. It would be naïve to imagine that school science could or should ever be the same as professional science. As one example, professional scientists may spend months or even years working virtually full-time on one problem, refining and repeating the same experiment, model or set of observations. School children need a breadth of science knowledge and, indeed, a broad curriculum of other subjects. It is important then to realise that, by its very nature, school science is going to be something other than science itself, so decisions need to be made about the ways in which school science can sensibly offer an authentic representation of what science actually is.

But what is the nature of science?

Science is an area of human activity that has evolved over a number of centuries, and there is no simple clear definition of what counts as 'science'. However, one core feature of science is that it involves an iterative process of developing knowledge by moving between empirical evidence and theory. That is, the collection of data supports the development of theories and models, which in turn suggest which data to seek next to refine and develop those theories and models. Although practical work is often a key feature of school science, it seldom gives students a strong feel for this ongoing relationship between theory development and testing.

One common suggestion is that an activity is science if it uses 'the' scientific method (which usually means an experimental method), but actually there is no simple method that applies across all the sciences. In some scientific fields, experimental manipulation is either inappropriate or just not possible (in much of astronomy, for example) and more naturalistic observational methods are needed, alongside the construction and testing of models. Even when experimental methods are used, it is seldom a matter of forming a hypothesis, designing a test for it, and simply carrying out the experiment.

Often there is a great deal of development of the design of an experiment, and the process may involve building, refining or calibrating new instruments. 'Science-in-the-making' often depends upon new theoretical thinking, new methods, and techniques that are still to be standardised. This makes it much less tidy than the image of science often presented in school texts: in particular, it relies upon imagination and creativity as well as logical argument. It also means that the scientists genuinely do not know what the findings will be (unlike many school practical 'investigations' where everyone knows what is meant to happen at the outset), making it potentially very exciting.

Much of the knowledge developed during the intense work of science is tacit (i.e. implicit knowledge), which is not available for discussion, so that scientists cannot fully justify their decisions – they are using a 'feel' for the situation and relying upon intuition (Brock, 2015). Scientific findings are then not simple logical deductions that must be true, as the interpretation of the data is always made within some theoretical framework, and usually relies upon complex data collection (drawing on practical knowledge that is partially tacit) and analysis techniques, and so upon judgements that cannot be fully rationalised. This may seem to undermine the popular model of science as producing objective knowledge based on readily-interpreted results from critical experiments, and makes science seem much more arbitrary and subjective. However, our confidence in science does not derive from definitive results from critical experiments carried out by geniuses (even if textbooks may suggest this), but rather from the iterative nature of scientific work – the cycle of building theory upon observation, and developing tests from theory – that is self-correcting over time.

Scientific thinking

Traditionally, science has been associated with objective rational thought and the kind of logic that is used in control of variables in investigations. Whilst this is certainly an important foundation for scientific thinking, real-life problems can be much more complex, and may need to be tackled without complete or entirely objective information, or may depend upon recognising value positions as well as examining technical evidence (Zeidler, 2014). It has been suggested that an authentic science education that reflects the nuanced nature of science, and especially one that includes consideration of socio-scientific issues (where science informs public policy without being the only consideration), may support cognitive development and, in particular, offer suitable challenge for those students sometimes labelled as 'gifted' in science (Taber, 2016).

Science *does* draw upon rational thinking that can apply logic to understand which options are consistent with, or ruled out by, particular observations. However, science also relies on creative thinking (Taber, 2011). Scientists have to imagine possible ways that the world is (structures, mechanisms, etc.) that may fit with available evidence, and imagine how these possibilities might best be tested. Sometimes this means creating imaginary features that help us to understand phenomena better (such as the imaginary lines of force drawn around magnets), or using analogies or metaphors as sources of new ideas.

Often in real world situations the information available is incomplete, or data are messy (perhaps including large experimental errors or erroneous readings), and judgement has to be used to decide how much inconsistency or incompleteness can be tolerated in drawing conclusions. When scientists are involved in making public policy decisions, they may also have to balance scientific evidence with other considerations (such as public opinion).

Scientific values and ethics

Scientific practices have changed over time, and will probably change further in the future. Science is a communal activity, where the international community (provisionally) judges the status of ideas through the process of peer review (determining which articles are published and which projects funded) and

Textbox 1: Some values commonly espoused in the scientific community

Science seeks objective knowledge – that is, knowledge that is in principle available to any member of the community. So, scientific work should be published so that it is available to other scientists, and should be reported in sufficient detail to allow others to repeat the work to check that they can replicate the findings.

Scientific work is judged on its merits without consideration of the status of the scientist(s) concerned. (It is well known that Einstein wrote a number of his key early papers in his spare time whilst working as a patent examiner. Being an unknown researcher without a scientific position did not prevent the merits of his work being recognised.) There is a peer review system whereby experts are asked to evaluate manuscripts submitted for publication in terms of their quality – often without being told who has written the paper.

Scientists should be unbiased in reporting their work – so report negative results as well as positive findings, and offer the best-evidenced (not the personally-preferred) interpretation of their data.

Science is an iterative community activity, where each contribution builds upon others. Research reports should acknowledge (cite) prior studies informing new work, as well as any colleagues offering material support (providing samples for example) or advice.

citations (so that some publications are much discussed and built on, whilst many are largely ignored). The community of practice that comprises contemporary natural scientists tends to share a number of principles that might be considered scientific values. Some key points are presented in Textbox 1 (above).

It is important to point out that this list is prescriptive (a set of ideals), not descriptive; occasionally, individuals do fall short. Scientists are humans who have been found to often be more wedded to their personal hunches than a detached objectivity would require. Moreover, journal referees are likely to see studies with positive results as more significant and worthy of journal space than those that offer more negative outcomes. Replication does not happen as much as the ideal would imply, as grants are more likely to be awarded for, and merit given to, new work.

Scientists often have strong commitments to particular theoretical ideas and methodological approaches, which have built up as they have invested heavily in becoming experts within a field. Extensive study of a particular research tradition leads to expertise that relies on a close familiarity with a particular way of thinking. Although this seems inconsistent with the notion of scientists being objective and neutral, and does sometimes impede progress, it is also inevitable. In a highly developed field of research, further advancement relies upon scientists having a detailed understanding of the current knowledge about a topic and the techniques that have proved productive (Kuhn, 1970). Modern science is usually communal and different scientists need to share commitments to aims, theory, definitions and methodology, in order to work together and to communicate effectively (which often means talking a kind of shorthand only well understood in the field). Consequently, scientists can sometimes disagree upon what data mean when they approach the evidence with different expertise developed in different scientific traditions.

Scientific ethics require scientists to adopt the values of objectivity and full reporting as best they can. Very occasionally, it has been found that scientists have fabricated, distorted, or deliberately selected favourable data in their studies. It seems that sometimes the lure of being right, or of making a breakthrough to bring fame (and likely better career opportunities) weighs more than the genuine desire to better understand nature in motivating most scientists. It is likely that the difficulties of being totally objective distort the research literature much more than deliberate fraud.

Scientific and extra-scientific commitments

As science is a human construction, and has changed over time, there are sometimes different views on which assumptions (sometimes called metaphysical assumptions) should be considered as core to scientific practice (see Table 1 overleaf). It is widely agreed that science investigates the material world, about which there can be, in principle, objective knowledge. Scientists therefore assume that there is a physical universe in which we live, which is to some extent knowable through observation and measurement. Scientists also assume (in part based on observations) that nature itself has sufficient regularity and stability that it is reasonable to look for general principles and laws that will be applicable everywhere, and at any time.

Several centuries ago, it was commonly assumed that science could produce fairly definitive knowledge, as human cognition was capable of producing objective knowledge (that is, knowledge that could be definitively agreed upon by different careful observers). Much more is now known about the biases and limitations of human perception and cognition, and the logical challenges of interpreting data, but it is still a common assumption in science that we can develop robust knowledge that helps us understand and explain (and sometimes control) natural phenomena.

Table 1: Some underlying assumptions adopted in, or sometimes associated with, science		
Metaphysical assumption	Status among scientists	Note
There is an external physical universe that is the same for all observers.	Generally adopted.	Science does not really make sense unless one makes this assumption.
The universe is at some fundamental level stable (so that law-like patterns may be observed).	Generally adopted.	Science does not really make sense unless one makes this assumption.
Humans are capable of understanding the natural world.	Generally adopted, but usually not in absolute terms.	Advances in cognitive science and philosophy of science suggest human understanding has limitations, undermining the confidence shown by some historical scientists.
Science can provide objective knowledge.	Generally adopted, although the degree to which such knowledge is limited and necessarily imperfect is a matter of discussion.	Some scientists judge scientific knowledge in pragmatic and instrumental rather than absolute terms (i.e. that it is good enough to be useful, even if incomplete).
Scientific processes lead to progress in understanding the world.	Generally adopted when taking a long-term view.	It is commonly recognised that progress may be punctuated by 'false starts' (and 'cul-de-sacs'). However, scientists reject 'relativist' views that science is purely a cultural product (rather than reflecting what is inherent in nature).
Science provides a method for developing objective knowledge.	Generally adopted in terms of the overall process, although there is no single 'scientific method'.	Science is a process that uses empirical observations to develop and test theory, which guides further data collection, but not always using experiments and control of variables.

Table 1 cont: Some underlying assumptions adopted in, or sometimes associated with, science

Metaphysical assumption	Status among scientists	Note
Science allows us to know the work of a creator God.	Not a commitment common to the scientific community.	A very common idea among scientists at one time, but now held as a *personal* belief by some scientists, without being seen as something inherent to science.
Science replaces belief in God or the supernatural.	Not a commitment common to the scientific community.	A personal view, or aspiration, of some individual scientists, but rejected by most, and not seen as inherent to science.
Everything can and will be explained by science.	Not a commitment common to the scientific community.	Some ('scientistic') scientists believe or hope this will happen. Some other scientists feel it may be possible in principle, but is unlikely in practice.
Science is the best way to explain all phenomena.	Not a commitment common to the scientific community.	Some ('scientistic') scientists take this view, but most scientists feel that there are many areas of human experience, which – even if in principle open to scientific explanations – are more productively understood in other ways.

Scientific knowledge is then always provisional in the sense that new evidence may lead to us reviewing our ideas (as when Newton's mechanics was superseded by Einstein's ideas). Some scientists may consider scientific knowledge to be good approximations to the true nature of the world (so Newton's mechanics gives very accurate predictions in nearly all applications, even if not a perfect account), and others may consider scientific knowledge just as an instrument, where we can be content as long as the theories and models work well enough for our purposes. Although there are differences in how scientists think about this, an authentic science education should make it clear that models and theories are human constructions – acts of creative imagination to fit available evidence – and that they should not be considered absolute descriptions of nature. Of course different ideas, in different fields, have somewhat different status. Some principles, laws and theories seem to do a very good job of accounting for all available relevant evidence. In other areas, scientists are aware of major limitations, suggesting that current models and theories are likely to need to be significantly refined or even replaced in the future. This should make science an attractive area of work to some students, as it still has much scope for major contributions.

Science and belief

It should be clear then that science is not something that should be believed in an absolute sense. It is not the job of the science teacher to tell students what to believe about nature – whether the topic is natural selection, the Big Bang, conservation of angular momentum, or d-level splitting – but (a) to understand these ideas and (b) to appreciate why they have been developed in relation to the available scientific evidence.

One area of contention in some science classes concerns the question of the supernatural – things that may exist that are in some sense outside of the natural world available to scientific investigation: for example, entities like ghosts, spirits, angels, and jinn. In some cultures, supernatural beliefs compete with scientific ideas. So, illness may be blamed on an evil spirit or a witch, rather than an infectious microorganism. This can have real consequences if the ill person refuses medicine and instead seeks some kind of exorcism, or, as continues to occur even in the 21st Century, children are killed because they are believed to be witches.

An issue that has received a great deal of publicity is the potential link between science and atheism (a belief that there is no God). A number of high profile scientists have suggested that there is no God or supernatural realm and that religion is an enemy of science and rationality. This is relevant to all science teachers (whatever their own views about religion), as students who have met this idea, and who have religious faith, may wonder if a career in science is incompatible with their beliefs.

It is not for a science teacher to tell students what to believe about nature, or about the supernatural. Rather, teachers should help students to understand why ideas are, or are not, considered to be supported by the available evidence. It is important though that students do understand that, by definition, natural science does not deal with a possible supernatural realm.

To appreciate this aspect of the nature of science, it may be useful to offer pupils some historical perspective. Many of the famous early modern scientists were people of strong faith (Galileo, Newton, Hooke, Kepler and Faraday, among others), and they often saw science as a means to better understand the works of their creator God. However, this does not mean that all scientists *today* must believe in God, any more than the publicity sought by some radical atheists today means that scientists should reject God. Indeed, the notion of being agnostic – that is, of considering that one cannot be certain of either the existence or non-existence of God – was developed to reflect the scientific attitude of remaining open-minded. Today, there are successful scientists around the world who are theists (believers), agnostics and atheists, as well as a great many who have not settled on a definitive position.

Atheism is sometimes associated with a position called scientism, which posits that not only does everything fall under the auspices of science (as there is only a natural realm), but also that science can and will one day explain everything. Although some scientists hold this view, the majority of scientists think that there are likely to be limits on science (i.e. on human understanding), yet science can still make much progress and usefully contribute to society in important ways.

In conclusion

This brief account suggests that the nature of science is (a) complex and nuanced; (b) fascinating; and (c) something that any authentic science education should acknowledge and reflect. We conclude by offering some advice to teachers:

➢ Teach scientific knowledge as human constructions (models, theories, etc.) that are open to revision, not as absolute accounts of nature;

➢ Encourage the scientific attitude: to be critical (including of one's own and widely accepted ideas), to look for alternative explanations, to be open to new evidence;

➢ Emphasise the complementary role of imagination and logic;

➢ Allow students to develop models, and various original kinds of representations, and to explore analogies and metaphors;

➢ Include genuine investigative work (i.e. that is really open-ended);

➢ Include naturalistic observations as well as experimental practical work;

➢ (At least sometimes) use practical work to motivate theory/model-building rather than to illustrate theory;

- ➤ Sometimes undertake extended practical work, especially where experimental design needs to be developed and refined to overcome initial limitations;
- ➤ Include vignettes of historical and contemporary scientific work to illustrate the power and limitations of science;
- ➤ Occasionally commit time to a detailed case study of the development of a scientific idea that can illustrate the complexity of science in action;
- ➤ Include examples of socio-scientific issues to highlight how scientific knowledge contributes to, but underdetermines, policy decisions that are also informed by value positions unrelated to the science; and
- ➤ Emphasise the cultural importance of science to modern life, whilst acknowledging that there are limits to our knowledge – science makes a major contribution, complementing other areas of human experience (such as aesthetic appreciation).

Further reading

i) Approaches and resource ideas:

Allchin, D. (2013) *Teaching the Nature of Science: Perspectives and Resources.* Saint Paul, Minnesota: SHiPS Educational Press

School Science Review (June 2006) Special issue on 'Ideas and Evidence', **87**, (321)

Taber, K.S. (2007) *Enriching School Science for the Gifted Learner.* London: Gatsby Science Enhancement Programme

ii) Introductions to research:

ASE Guide to Research in Science Education (new edition forthcoming in 2018/19).

References

Brock, R. (2015) 'Intuition and insight: two concepts that illuminate the tacit in science education', *Studies in Science Education*, **51**, (2), 127–167. DoI:10.1080/03057267.2015.1049843

Kuhn, T.S. (1970) *The Structure of Scientific Revolutions (2nd Edition)*. Chicago: University of Chicago

Taber, K.S. (2011) 'The natures of scientific thinking: creativity as the handmaiden to logic in the development of public and personal knowledge'. In *Advances in the Nature of Science Research – Concepts and Methodologies*, Khine, M.S. (Ed.), (pps. 51–74). Dordrecht: Springer

Taber, K.S. (2016) 'The nature of science and the teaching of gifted learners'. In *International Perspectives on Science Education for the Gifted: Key issues and challenges*, Taber, K.S. & Sumida, M. (Eds.), (pps. 94–105). Abingdon: Routledge

Zeidler, D.L. (2014) 'Socioscientific Issues as a Curriculum Emphasis: Theory, research, and practice'. In *Handbook of Research on Science Education*, Lederman, N.G. & Abell, S.K. (Eds.), 2, pps. 697–726). New York: Routledge

CHAPTER 3

Educational neuroscience and the brain: some implications for our understanding of learning and teaching

Derek Bell & Helen Darlington

The question of what education can learn from neuroscience is a pertinent one: this chapter explores the field of educational neuroscience and the insights it offers about teaching and learning, along with the limitations and risks of this knowledge, including common 'neuromyths'. Much of the research supports the good teaching practices described later in the book, and should encourage teachers to explore these further.

Introduction

The idea that a greater understanding of how the brain works can improve teaching and therefore learning is very seductive, but what can we, as teachers and other professionals working in education, learn from neuroscience? We don't have to look far to see that the link is not straightforward and that there is no 'silver bullet'. If we proceed with caution, however, there is much to be gained by reflecting on the insights that are now forthcoming from research that crosses the boundaries of neuroscience, cognitive sciences, psychology and education. It is evidence from work in this multidisciplinary field of 'educational neuroscience' that we draw upon in writing this chapter. Our starting point is to reflect briefly on the growth of activity in educational neuroscience, its potential, risks and limitations, including examples of what are often referred to as neuromyths. We then consider how the brain works before exploring some examples of learning processes and, importantly, the implications that these might have for teaching.

The growth of educational neuroscience

The vast majority of our understanding of learning is based on behavioural studies, which have shown relationships between interventions, children's responses and their learning. These findings in turn have led to a wide range of theories of learning. Some of these have stood the test of time and/or have been associated with models of how particular aspects of learning are processed in the brain. It is only recently, however, that it has been possible to begin to relate behavioural outcomes to the neural mechanisms that underpin them.

Advances in understanding how the brain works have gone hand in hand with developments in technology. Early studies of the brain involved careful, systematic observations of individuals who, for various reasons, had suffered brain damage or undergone surgery that resulted in changes in behaviour; Phineas Gage and Henry Molaison[1] are perhaps the best known examples. For obvious reasons, the number of such studies is limited and so it wasn't until the introduction of different brain scanning techniques that it was possible to even consider monitoring changes in the brain while individuals were carrying out specific types of task. Although fMRI (functional magnetic resonance imaging) scans, with their pictures of brains showing 'hotspots' of colour, are perhaps the most familiar, there are others, e.g. EEG (electroencephalography) and associated ERP (event-related potentials), which together have provided insights into which parts of the brain are involved in different processes (Dick *et al*, 2014). In addition, increased emphasis on computational models of neural networks has furthered our understanding of the links between neural networks and the behaviours that they support (Thomas & Laurillard, 2014). Thus there are more tools at our disposal, but the potential of educational neuroscience is that it adopts a more integrated approach, bringing together not only the findings related to the structure and function of the brain with those from psychology, but also evidence from what happens in classrooms and other learning environments.

Neuromyths – some common misconceptions

As with any good research, caution is required to avoid jumping to conclusions or extrapolating findings to a point beyond which there is no supporting evidence. Taking 'the latest research findings' and applying them to learning without careful thought runs the risk of oversimplification, misinterpretation or simply being inappropriate. Indeed, such misuse of evidence, combined in some cases with over-promotion, has led to the spread of neuromyths. A more positive effect has been the increasing enthusiasm and belief of teachers and others that a better understanding of brain function is relevant to the way in which they think about learning and approach their teaching.

While some claims are demonstrably wrong, it is not so easy to dismiss others outright. In part, this is because there is a glimmer of scientific fact, which suggests that there might be some truth in the claims despite the lack of supporting evidence. 'Learning styles', a concept widely adopted by teachers 5 to 10 years ago but now declining, is a high profile example of how a neuromyth can develop. Based on the fact that processing of visual, auditory

[1] For further details see: https://digest.bps.org.uk/2015/11/27/psychologys-10-greatest-case-studies-digested/ (accessed June 2017)

and sensory stimuli is associated with different regions of the outer layer of the brain (the cortex), it was suggested that some children responded more readily to information presented in a particular way – their 'preferred learning style'. Thus it was argued that, in order to maximise their learning, children should be taught using their preferred learning style. This is a very attractive idea but, despite studies to test this hypothesis, it is not supported by the evidence; on the contrary, multi-sensory approaches to learning are more effective.

As interest in educational neuroscience continues to grow and the risk of neuromyths increases, it cannot be emphasised too strongly that, as in all areas of science, we have to consider not only all the evidence available but also the quality and robustness of that evidence. At best, neuromyths are misleading but, more fundamentally, they undermine trust in claims that are genuinely underpinned by sound evidence. Table 1 summarises some of the most common neuromyths, but also draws attention to some ideas where the evidence is more supportive and which, on balance, might be considered to be 'neurohits'.

How the brain works: structure, development and function

The human brain is made up of two main cell types: glial cells and neurons. For over 50 years there were thought to be 1 trillion glia and 100 million neurons. However, improved counting methods suggest that this is itself a neuromyth and that the numbers are much smaller, more in the region of 85 billion glial cells and 86 billion neurons. Although the role of glial cells is under-researched in comparison with neurons, it is clear that they complement each other in the overall functions of the brain. Glial cells do not transmit impulses, but perform a range of roles, including forming the myelin sheaths that develop round the axons, providing physical support for the axons, digesting dead cells and helping to regulate the contents of the extra-cellular space. In contrast, neurons, each with their dendrites and axon, receive and transmit nerve impulses in the brain, thus directing how we react to environmental stimuli. At this cellular level, neurons are not connected to each other directly, but at synapses where the ends of axons and dendrites come close together. Impulses are then transmitted chemically from one neuron to the next (see Figure 1 p.30). The more frequently such connections are used, the more robust and reliable they become. Such effects might relate to physical skills such as hitting a tennis ball, or recalling information in order to answer a quiz question or solve a problem. The potential for making connections is enormous, which is why humans have such a capacity to learn new things and adapt to new experiences and situations throughout their lives. The ability of the brain to renew and make new connections is referred to as 'brain plasticity'.

Table 1: Common neuromyths and neurohits[2]

We only use 10% of our brains	MYTH – essentially we use all of our brains all of the time (though not all neurons are firing at the same time).
Left brain versus right brain thinkers	MYTH – no evidence to support the notion that people are intrinsically left- or right-sided in how their brains are wired, or that hemispheric specialisation has any relevance to individual differences in learning ability.
Girls and boys have different cognitive abilities	MYTH – there are differences, but they are smaller than once thought. Any differences that do exist are certainly not relevant to academic potential.
Different children have different learning styles	MYTH – as much as the idea is intuitively plausible, there's no good evidence to support the use of teaching through the preferred learning styles.
Most learning happens in the first 3 years of life	HIT – this is a neurohit in that the first three years are a vital period of brain development, but a MYTH – in that the only aspect of higher cognitive development that is really dependent on those early years is social and emotional security.
Physical exercise enhances learning	HIT – aerobic exercise appears to be effective in improving cognitive performance in children through boosting brain plasticity.
Well-rested children do better in school	HIT – in general, well-rested children will perform better in the classroom in the short term and the long term.
Diet makes a difference to learning	HIT – lots of work to be done on which aspects of diet influence what sort of thinking, but good nutrition in the first two years of life is crucial.
Mindfulness has a place in the classroom	HIT –verdict isn't wholly clear but, given the emerging evidence, the theory looks good. Some claims are grander than the evidence supporting them and more carefully controlled studies are required.

[2] Table 1 has been compiled from a series of articles produced by the Centre for Educational Neuroscience, London (a university-led research centre across University College London, Birkbeck University of London and UCL Institute of Education), which can be accessed at: http://www.educationalneuroscience.org.uk/neuromyth-or-neurofact/

Figure 1: Neuron structure

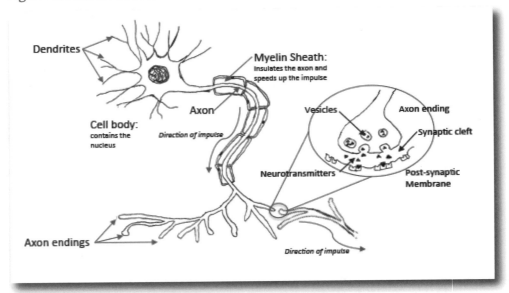

In the first few years of life, the brain is making large numbers of connections, a process called synaptogenesis, as newborn infants interact with the myriad of sights, sounds, smells and other stimuli around them. At the age of 2 years, there are estimated to be 100 trillion synapses, approximately twice as many as in the adult brain. The reduction in number is because, as brains develop, connections that are frequently used become reinforced, but those that are not used are broken and lost. This process of synaptic pruning takes place at different developmental stages throughout childhood and adolescence. In addition, usually from puberty onwards, the efficiency with which impulses are transmitted is improved by the growth of myelin sheaths around the axons. This process of myelination is particularly active during adolescence when other major changes in the brain are also taking place, most notably the maturing of that part known as the pre-frontal cortex (PFC). This is the last part of the brain to develop, maturing around the age of 25, and is responsible for a series of cognitive processes, referred to as executive functions, and aspects of complex behaviour including assessment of risk. It is these changes along with those in hormone levels that result in 'teenager behaviour': the desire for quick rewards, a tendency to take risks especially in the presence of peers, emotional highs and lows, swings in attitudes towards learning and changes in biological rhythms and sleep patterns. In adulthood, when unaffected by conditions such as dementia, the number of neurons tends to be stable, but changes in the number of connections continue throughout life especially when the individuals are active both mentally and physically.

Figure 2a: Structure of the brain – external features

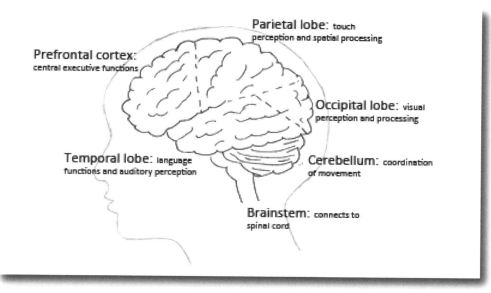

Figure 2b: Structure of the brain – cross-section

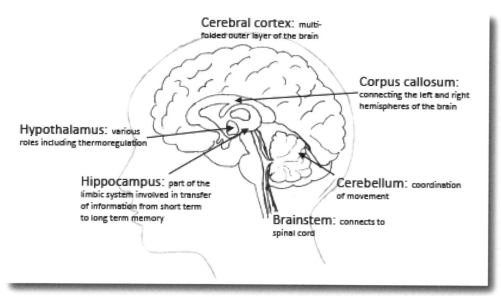

To fully appreciate how the brain works, it is necessary to consider not only the detail of the cells and the stages of development, but also how the brain functions as an organ in controlling our actions, behaviours, learning and emotions. We know it consists of two halves, which are firmly joined by the corpus callosum, and that it is layered. Most of our understanding comes from the study of the folded outer layer, the cortex. Its different areas, known as lobes (see Figure 2a), are involved in processing the various types of information. It is important to note, however, that other parts that have key roles in learning, memory and emotions, e.g. the limbic system including the hippocampus, are underneath the cortex, nearer the centre of the brain (see Figure 2b). Brain activity ultimately builds on the networks that result from the multiple connections that are formed not just between individual neurons, but also between groups of neurons in different parts of the brain. Particular regions may be involved in multiple networks and so contribute to processing a wider range of different activities. The crucial point is that, unlike a computer, which generally handles data sequentially through a central processing unit, the brain is involved in extensive parallel processing of information and pattern recognition.

Implications for teaching and learning

Conceiving the brain as a parallel processing organ, which establishes neural networks, has implications for understanding learning and for teaching. For example, it suggests that learning is fundamentally domain-specific because connections are made within particular contexts. Unless there are deliberate steps taken to help children make links across subjects and situations, the transfer of knowledge and generalisation of underlying principles is less likely to happen. Not surprisingly, this is one of the items in Table 2, which summarises some key features of learning and possible implications for teaching. Looking at this Table, the first thing to note is that the evidence from educational neuroscience in general supports many of the features that might be described as 'good practice' in learning. The second point is that, looking behind the headlines, it challenges us to ask how effectively these features are catered for in our teaching.

More specifically, but not exclusively, learning in science involves establishing knowledge (building a bank of evidence), understanding processes (generation of evidence) and developing concepts (understanding phenomena and their relationships) (Tolmie, 2016). By bringing these three aspects together, children begin to investigate and make sense of the world around them in scientific terms. To achieve this, teaching science requires pupils, regardless of age, to be engaged in experiences and activities that:

➢ involve them in manipulating causal events;
➢ direct attention to the exact sequences and components that make up the event or structure;

> link descriptions and explanations explicitly to the different elements of the events; and

> connect the experiences to explanatory constructs and the bigger conceptual ideas of science.

Table 2: Features of learning and examples of implications for teaching

Features of learning	Example implications for teaching
Importance of language and social interaction	Are children encouraged to use appropriate vocabulary in the correct contexts? What opportunities are there for children to share ideas, formulate questions and draw conclusions?
Use of multi-sensory approaches	To what extent do practical activities involve all the senses? Are children challenged to provide different forms of evidence in support of their ideas?
Range of experiences	How many instances of a process (e.g. evaporation) do children explicitly experience?
Influence of pre-existing ideas	Are children encouraged to be explicit about their pre-existing ideas and examine them against alternative explanations?
Need for scaffolding	What types of support are used to help children extend their understanding of phenomena?
Making things explicit especially if transfer of learning is to be supported	To what extent are children encouraged to demonstrate links between cause and effect? In what ways are linking ideas (e.g. classifying objects and organisms) across topics and subjects made explicit?
Quality of the learning environment	Does the physical environment provide stimuli that encourage children to ask questions/ reinforce ideas? Do children feel safe offering explanations, even if they turn out to be incorrect?

Learning in science, as in other subjects, depends on the use of the executive functions in context. These are the underlying processes responsible for our ability to direct, maintain and focus attention, manage impulses, self-regulate behaviour and emotion, plan ahead and demonstrate flexible approaches to problem-solving (Howard-Jones, 2010 p.190). As executive functions are mainly associated with the pre-frontal cortex, the last part of the brain to fully develop, it does not mean that young children are unable to use such skills. On the contrary, they can, but need more support and experience in doing so. We now consider three aspects of learning related to the executive functions and some implications for teaching.

Reasoning and inhibition

Reasoning is a key element in learning in science and requires linking together not only ideas but also linking perceptual observations to existing ideas, in order to interpret the evidence in relation to scientific hypotheses and theories. To do this often requires the suppression of prior knowledge and intuitive ideas, which are in conflict with new evidence and possible alternative explanations. This process, which is known as inhibitory control, or inhibition, is a key element in cognitive development. This is consistent with evidence from behavioural studies that indicates that prior (intuitive) ideas are not erased from the brain once something new is learnt (Shtulman & Valcarcel, 2012). Indeed, the evidence from brain imaging studies indicates that different neural pathways are used depending on whether the reasoning is consistent or inconsistent with existing prior knowledge. Perhaps more significantly, when faced with counter-intuitive situations, experts in the subject show more brain activity in those areas associated with inhibition than do novices (Masson et al, 2014). This further underlines the importance of the inhibitory function, which predominantly involves two areas of the brain anterior cingulate cortex (ACC – identifies the inconsistency) and the dorsal lateral pre-frontal cortex (DLPFC – attempts to resolve the problem). Resolving the conflict may involve 'seeing' alternative interpretations of the evidence, and/or innovative solutions to problems, a process sometimes referred to as cognitive flexibility. Support for improving reasoning and inhibition not only involves practice, but also the challenge to ideas and the robustness of arguments. Both require the use of strategies that encourage thinking, such as:

➢ allowing more time (referred to variously as 'wait-time', 'processing time', 'thinking time') for children to answer questions, i.e. giving them 3-5 seconds rather than the common practice of less than 1;

➢ making the 'stop and think' process explicit and asking for reasons to justify answers;

> ➤ specifically checking children's existing ideas, taking them seriously, testing them and explicitly relating them to the new idea – this can productively result in what is referred to cognitive conflict;
> ➤ encouraging consideration of alternative ideas/explanations, with justifications, using, for example, concept cartoons;
> ➤ providing opportunities to express ideas in a range of forms, e.g. verbally, in writing, pictorially and physical modelling; and
> ➤ using 'pair-share' activities to give children the opportunity to consider their ideas with peers before 'going public'.

Memory

Memory, both short- and long-term, is central to learning through the encoding, storage and retrieval of information and ideas. Long-term memory stores large amounts of information that is retained by the established connections linking new with existing knowledge. Depending on the type of information, it is held in different parts of the brain and brought to bear on a problem when needed. 'Memories', as such, are not recalled as one might pull out a photograph of a landscape or event. Rather, they are reformed by bringing together information from the various locations and so our recollections are not exactly the same on each occasion, but become more reliable the more frequently that 'memory' is used. In order to become encoded, we need to attach meaning to the information. In terms of brain function, this involves linkages with a wider set of neural networks that can mutually support the encoding and retrieval of the information. We also need to practise the function in a range of contexts, whether we are trying to improve the way we perform physical actions, e.g. kicking a ball, or our ability to recall particular types of information. The more secure information is in long-term memory, the more efficient its retrieval and the less energy is used in 'thinking'. This in turn leaves more energy available to think about and learn new material or find solutions to problems.

The role of short-term memory, or working memory, involves handling the range of information required to carry out tasks in which we are engaged; this may be simply holding a list of instructions that are to be followed or maintaining the answers to a series of operations in a mathematical calculation. Three important things need to be noted: the first is that the amount of information that can be held in the working memory is limited (2 to 4 chunks for young children, increasing to 5 to 7 for adolescents and adults). The second is that, once the information has gone from the working memory, it cannot be retrieved unless it has been transferred successfully into the long-term memory.

The third is that, if the working memory has reached its capacity, then overload may occur thus inhibiting further progress with the learning in question. For young children, poor working memory may mean they are unable to follow a set of instructions and so end up wandering around the classroom not knowing what to do next (see Gathercole & Alloway, 2008). For adolescents, it may be they cannot relate evidence gathered in a practical lesson to the theoretical explanation of the concepts involved. Having reliable knowledge in the long-term memory can reduce the overload and make new learning more likely. Many of the problems can be tackled by modifying teaching strategies accordingly, for example:

> ➢ providing fewer instructions and/or giving written or pictorial reminders of the actions required;
> ➢ working in mixed-ability groups so that the information can be shared and children support each other;
> ➢ supporting the development of working memory over time by increasing the demand, but note that this is not unlimited;
> ➢ encouraging children to develop their own strategies, e.g. listing steps required, use of 'working notes' and flow diagrams; and
> ➢ using complementary 'input' sources, e.g. visual (diagrams/pictures) and phonological (text and verbal instructions).

Motivation, interest, emotion, anxiety and stress

Learning is not based solely on cognitive abilities, but involves a significant affective component. If children are not emotionally engaged with the content, their learning is likely to suffer. There is no question that severe stress and anxiety affect the way in which the brain functions and that in turn impacts on learning. Children who have experienced chronic stress as a result of, for example, physical abuse or early neglect, have a smaller hippocampus, correlating with deficits in short-term memory, learning and ability to manage future adverse events (Hanson et al, 2015). More commonly, changes in the balance of the brain's chemical messengers (neurotransmitters), such as noradrenaline and dopamine, interfere with the operation of various areas of the cortex, resulting in psychological symptoms (e.g. depression or anxiety). In addition, when students experience a perceived stressor, the hypothalamus will trigger the 'fight or flight' response, leading to increased heart and breathing rates, among other things. Neither of these physical or psychological effects is conducive to learning. Some children exhibit forms of anxiety specific to an area of learning; the occurrence of maths anxiety, for example, is well documented (Hill et al, 2016). Although some of the causes of anxiety are sometimes beyond the control of teachers, they must be sensitive to the possibility that external events will impact on a child's progress and attitude to learning.

More positively, low levels of stress resulting in smaller changes in the hormone balance can be beneficial, generating excitement, motivation and a sense of achievement. Although there is much still to be done in order to understand the neural pathways involved, triggering interest, curiosity and motivation is possible through the judicious use of external and internal motivators. However, reliance on external rewards becomes self-defeating. Furthermore, approaches that support children in developing behaviours to encourage a positive attitude to learning based on mindfulness (e.g. Zenner *et al*, 2014) and metacognition (see Hattie, 2009) appear to have an impact in terms of children's self-awareness and their progress. Teaching strategies that provide opportunities for such development include:

- ➢ effective use of group work in which children are able to provide mutual support and learn from each other;
- ➢ creating an environment in which children feel they have some level of control over what and how they approach a problem, e.g. practical work that involves them in deciding what they are going to investigate as opposed to simply following a set of instructions;
- ➢ an explicitly shared understanding of the purpose of both individual activities and the subject concerned;
- ➢ opportunities for children to make judgements about the quality of their work and improve it; and
- ➢ scaffolding work to vary the challenge involved, including the setting of achievable goals; this may also involve some choice for children as to which level of challenge they wish to take on.

In conclusion

Without question, our understanding of how the brain works has progressed greatly in recent years, but its complexity means that we are only scratching the surface. However, by bringing together the new evidence of neural pathways with behavioural data from other disciplines, including psychology, our understanding of learning is also progressing. The question is how these new insights might be applied effectively to teaching. Much of the evidence from educational neuroscience supports what we would describe as 'good practice', but it goes further and indicates the need to be much more aware of what appears to be going on in the brain and explicitly modify teaching approaches accordingly. Currently the evidence available is not clear or robust enough to suggest wholesale changes to curricula or practice, but we hope that, in this chapter, we have shown that there are subtle but significant and sustainable changes that can be made by us as teachers, both individually and collectively, in order to improve the learning of our children.

Further reading

Howard-Jones, P. (2010) *Introducing neuroeducational research: neuroscience, education and the brain from contexts to practice.* Abingdon: Routledge

Mareschal, D., Butterworth, B. & Tolmie, A. (Eds.) (2014*) Educational Neuroscience.* Chichester, UK: John Wiley & Sons

Neuromyths and neurohits. A series of short articles produced by the Centre for Educational Neuroscience, London (a university-led research centre across University College London, Birkbeck University of London and UCL Institute of Education) available at: http://www.educational neuroscience.org.uk/neuromyth-or-neurofact/ (accessed June 2017)

References:

Dick, F., Lloyd-Fox, S., Blasi, A., Elwell, C. & Mills, D. (2014) 'Neuroimaging methods'. In *Educational Neuroscience,* Mareschal, D., Butterworth, B. & Tolmie, A. (Eds.). Chichester, UK: John Wiley & Sons

Gathercole, S.E. & Alloway, T.P. (2008) *Working memory and learning: A practical guide.* Los Angeles, CA: Sage Publications

Hanson, J.L., Nacewicz, B.M., Sutterer, M.J., Cayo, A.A., Schaefer, S.M., Rudolph, K.D., Shirtcliff, E.A., Pollak, S.D. & Davidson, R.J. (2015) 'Behavior problems after early life stress: Contributions of the hippocampus and amygdala', *Biological Psychiatry,* **77,** (4), 314–23. Pdf available at: http://centerhealthyminds.org/assets/files-publications/HansonBehavioralBiolPsych.pdf (accessed June 2017)

Hattie, J.A.C. (2009) *Visible learning.* Abingdon: Routledge

Hill, F., Mammarella, I.C., Devine, A., Caviola, S., Passolunghi, M.C. & Szűcs, D. (2016) 'Maths anxiety in primary and secondary school children: Gender differences, developmental changes and anxiety specificity', *Learning and Individual Differences,* (48), 45–53. Pdf available at https://www.researchgate.net/profile/Maria_Chiara_Passolunghi/publication /297595498_Maths_anxiety_in_primary_and_secondary_school_childrens_G ender_differences_developmental_changes_and_anxiety_specificity/links/5 710b9f908ae74cb7d9f8a48.pdf (accessed June 2017)

Howard-Jones, P. (2010) *Introducing neuroeducational research: neuroscience, education and the brain from contexts to practice.* Abingdon: Routledge

Masson, S., Potvin, P., Riopel, M. & Brault Foisy, L-M. (2014) 'Differences in brain activation between novices and experts in science during a task involving a common misconception in electricity', *Mind, Brain and Education,* **8,** (1), 44–55

Shtulman, A. & Valcarcel, J. (2012) 'Scientific knowledge suppresses but does not supplant earlier intuitions', *Cognition,* (124), 209–215. Pdf available at: https://sites.oxy.edu/shtulman/documents/2012b.pdf (accessed June 2017)

Thomas, M.S.C. & Laurillard, D. (2014) 'Computational modelling of learning and teaching'. In *Educational Neuroscience,* Mareschal, D., Butterworth, B. & Tolmie, A. (Eds.). Chichester, UK: John Wiley & Sons

Tolmie, A. (2016) 'Educational Neuroscience and Learning'. In *The Sage Handbook of Curriculum, Pedagogy and Assessment, Volume 1,* Wyse, D., Hayward, L. & Pandya, J. (Eds.). London: Sage

Zenner, C., Herrnleben-Kurz, S. & Walach, H. (2014) 'Mindfulness-based interventions in schools – a systematic review and meta-analysis', *Frontiers in Psychology,* DoI: 10.3389/fpsyg.2014.00603

CHAPTER 4

Primary science and the implications for secondary science teachers

Helen Wilson

For many children, primary school is where they first encounter science lessons and the foundations of many of the 'Big Ideas' in science. Understanding these prior experiences and some of the challenges faced by pupils when moving to secondary school is vital for secondary school science teachers, as they continue to develop both the pupils' knowledge and understanding of science, and their enjoyment of science at school.

Introduction

Pupils' first experiences of science are important, not least because attitudes to the subject are formed early. In a report to the European Commission, it was highlighted that *'science teaching at primary school has a strong long-term impact. Primary school corresponds to the time of construction of intrinsic motivation, associated with long-lasting effects, it is the time when children have a strong sense of natural curiosity and it is the right time to tackle gendered patterns'* (Rocard, 2007, p.11).

Even the youngest pupils in primary school do not come to science lessons with 'empty' heads; they have many existing ideas about how the world works. Of course, it may or may not be the case that these existing ideas are in line with our currently accepted scientific understanding. Teaching is a complex skill and this includes assisting pupils to assimilate new information and experiences, relate them to what they already know (or what they think they know) and adjust their understanding accordingly. This is at the very heart of learning in science.

NEW EXPERIENCES ⟷ EXISTING IDEAS

Learning then takes place at the interface between the new experiences, which may or may not be practical, and the pupils' existing ideas. Teaching goes well beyond pouring knowledge into previously empty heads – it would be so easy if that was all there was to it.

Giving pupils opportunities to reveal their existing understanding is therefore essential in order to take their learning forward in science. Such elicitation allows pupils to reveal their misconceptions, but also to show their correct understanding, and I have been humbled to discover the depths of thinking of which some young pupils are capable. One example is a ten year-old pupil's response to my question, *'How do you know that the Earth is a sphere?'*: *'Because gravity comes from the centre of the Earth, because a sphere is the smallest shape you can make from the centre, it would most likely be pulled up into a sphere.'*

Lewis and Smith (1993) define higher order thinking as something that *'occurs when a person takes new information and information stored in memory and interrelates and/or rearranges and extends this information to achieve a purpose or find possible answers in perplexing situations'* (p. 136). This ten year-old had done just that: he pondered the question, related it to what he already knew about gravity pulling to the centre of the Earth and then reasoned that, if it pulled equally in all directions, the only shape possible is a sphere, because it is the only shape where every point on the surface is the same distance from the centre.

Whilst not every pupil who arrives in a secondary science classroom will be able to articulate their ideas in such a clear way, they most certainly will not have 'empty' heads about the subject. They have been learning science throughout their time in primary school – science is done in primary school and often done very well. However, there are key differences between primary and secondary science. Primary school teachers in England are normally considered to be generalists, with very few being subject specialists in science. The Royal Society (2010) report showed that, at primary school level, only 3% of teachers hold a specialist degree in science accompanied by an Initial Teacher Education (ITE) qualification. A study by the Wellcome Trust (2013) identified two broad categories of science leadership in primary schools. The first is *'the "class teacher" model, in which an individual leads and champions science across the school, and all class teachers teach science to their own children; and the second is the "science teacher" model, in which the science teacher is likely to have a science-based degree and teaches all the science within the school'* (p.1). Their online survey of 209 schools showed that the vast majority (95%) adopted the class teacher model, so, on the whole, all primary teachers teach science as part of their normal classroom practice.

However, primary teachers are specialists in pedagogy – how to teach – and this leads to some very creative primary science lessons. Recent research by the Office for Standards in Education (Ofsted) has shown that being a generalist is no barrier to good primary science teaching: 'in both *Science and Mathematics, the international averages show the highest achievement for Year 5 [age 10] pupils taught by teachers with a primary teaching qualification but no subject specialisation*' (Ofsted, 2016a p.45).

As at Key Stage 3 (age 11-14), the programmes of study in the National Curriculum for Key Stages 1 and 2 (ages 5-7 and 7-11) include 'working scientifically', and the lack of laboratories in most primary schools is no barrier to them undertaking good quality practical science investigations. The National Curriculum states that, during Years 5 and 6 (ages 10 and 11), pupils should be taught to use the following practical scientific methods, processes and skills:

➤ planning different types of scientific inquiries to answer questions, including recognising and controlling variables where necessary;

➤ taking measurements, using a range of scientific equipment, with increasing accuracy and precision, taking repeat readings when appropriate;

➤ recording data and results of increasing complexity using scientific diagrams and labels, classification keys, tables, scatter graphs, bar and line graphs;

➤ using test results to make predictions to set up further comparative and fair tests;

➤ reporting and presenting findings from inquiries, including conclusions, causal relationships and explanations of and a degree of trust in results, in oral and written forms such as displays and other presentations; and

➤ identifying scientific evidence that has been used to support or refute ideas or arguments.

This is a comprehensive set of skills that they should be bringing to secondary school – they are by no means coming with empty heads. By Year 6, pupils are expected to be able to plan and carry out a range of types of practical investigations, *including* 'observing over time; pattern seeking; identifying, classifying and grouping; comparative and fair testing (controlled investigations); and researching using secondary sources' ('*including*' indicates that this is not intended to be an exhaustive list). Therefore, by the end of primary, it is expected that they will be capable of designing and carrying out investigations that they have planned themselves and understand how to control variables when necessary. The notes and guidance in the National Curriculum state that:

'They should make their own decisions about what observations to make, what measurements to use and how long to make them for, and whether to repeat them; choose the most appropriate equipment to make measurements and explain how to use it accurately. They should decide how to record data from a choice of familiar approaches; look for different causal relationships in their data and identify evidence that refutes or supports their ideas.'

Primary teachers are adept at enabling pupils to undertake practical work that facilitates these skills, without the need for laboratory space. For example:

Observing over time	⟶	What happens to the shadow of the netball post over a day?
Pattern seeking	⟶	Do people with the longest legs jump the furthest?
Fair test	⟶	What affects the rate at which ice cubes melt? How is pulse rate affected by exercise?
Problem solving	⟶	Make an alarm that warns when the hamster cage is opened.

It is stated that the pupils should be *'taught the practical scientific methods, processes and skills through the teaching of the programme of study content'.* Unlike at Key Stage 3, the content areas are not formally divided up into biology, chemistry and physics, so primary school pupils are unlikely to know within which of these three disciplines they are working at any given point in time – they will know it all as 'science'. Taking just their last two years in primary school as a snapshot, the tables overleaf show the content areas in the programmes of study, reorganised into the different disciplines.

The Programmes of Study (PoS) are set out year by year in the National Curriculum, but it is not statutory when a school covers which sections, as long as all are completed by end of the key stage, so schools do have the flexibility

Year 5

Biology	Chemistry	Physics
Living things and their habitats: Describe the differences in the life cycles of a mammal, an amphibian, an insect and a bird. Describe the life process of reproduction in some plants and animals. **Animals, including humans:** Describe the changes as humans develop to old age.	**Properties and changes of materials:** Compare and group together everyday materials on the basis of their properties, including their hardness, solubility, transparency, conductivity (electrical and thermal), and response to magnets. Know that some materials will dissolve in liquid to form a solution, and describe how to recover a substance from a solution. Use knowledge of solids, liquids and gases to decide how mixtures might be separated, including through filtering, sieving and evaporating. Give reasons, based on evidence from comparative and fair tests, for the particular uses of everyday materials, including metals, wood and plastic. Demonstrate that dissolving, mixing and changes of state are reversible changes. Explain that some changes result in the formation of new materials, and that this kind of change is not usually reversible, including changes associated with burning and the action of acid on bicarbonate of soda.	**Earth and space:** Describe the movement of the Earth, and other planets, relative to the Sun in the solar system. Describe the movement of the Moon relative to the Earth. Describe the Sun, Earth and Moon as approximately spherical bodies. Use the idea of the Earth's rotation to explain day and night and the apparent movement of the Sun across the sky. **Forces:** Explain that unsupported objects fall towards the Earth because of the force of gravity acting between the Earth and the falling object. Identify the effects of air resistance, water resistance and friction that act between moving surfaces. Recognise that some mechanisms, including levers, pulleys and gears, allow a smaller force to have a greater effect.

Year 6

Biology	Chemistry	Physics
Living things and their habitats: Describe how living things are classified into broad groups according to common observable characteristics and based on similarities and differences, including microorganisms, plants and animals. Give reasons for classifying plants and animals based on specific characteristics. **Animals including humans:** Identify and name the main parts of the human circulatory system, and describe the functions of the heart, blood vessels and blood. Recognise the impact of diet, exercise, drugs and lifestyle on the way their bodies function. Describe the ways in which nutrients and water are transported within animals, including humans. **Evolution and inheritance:** Recognise that living things have changed over time and that fossils provide information about living things that inhabited the Earth millions of years ago. Recognise that living things produce offspring of the same kind, but normally offspring vary and are not identical to their parents. Identify how animals and plants are adapted to suit their environment in different ways and that adaptation may lead to evolution.	None	**Light:** Recognise that light appears to travel in straight lines. Use the idea that light travels in straight lines to explain that objects are seen because they give out or reflect light into the eye. Explain that we see things because light travels from light sources to our eyes or from light sources to objects and then to our eyes. Use the idea that light travels in straight lines to explain why shadows have the same shape as the objects that cast them. **Electricity:** Associate the brightness of a lamp or the volume of a buzzer with the number and voltage of cells used in the circuit. Compare and give reasons for variations in how components function, including the brightness of bulbs, the loudness of buzzers and the on/off position of switches. Use recognised symbols when representing a simple circuit in a diagram.

to introduce content earlier or later. However, it is likely that most schools will address the areas in the order suggested. This latest National Curriculum is not designed to be as spiral in nature as the previous version, so concepts are not necessarily revisited. Sound, for example, is only included in the Year 4 (age 9) PoS and, whilst it has received attention in Year 5, there is no chemistry in the Year 6 PoS, so these areas may not be very fresh in the pupils' minds when they come to secondary.

Overall there is less physics content than in the previous curriculum and none at all in Key Stage 1. The logic of this seems to have been that the concepts were too challenging for young children. It could be suggested, of course, that it is preferable to meet such areas early in school so that the basics, such as forces being pushes and pulls, are embedded early in a pupil's understanding.

Transition

An effective transition between schools is obviously very important. Ofsted (2013, p. 50) make the point that *the best transition ensures that pupils get off to a flying start in their academic studies at secondary school'*. However, this is by no means simple and there are a number of potential issues.

Year 7 (age 12) pupils will certainly arrive with experience in science as a subject. However, any secondary school is likely to receive Year 7 pupils from a large number of primary schools and there may be a wide variation in these pupils' experiences. There were many reasons for the removal of the Key Stage 2 Science SAT in 2009 and amongst them was the hope that, as schools would feel less inclined to 'teach to the test', they would be free to be more creative in their science teaching. However, this has had an impact on the prioritisation of the subject. Ofsted (2013) in their survey into science education in schools noted that *'most [primary schools] prioritised English and mathematics above science, which is still a core subject in the National Curriculum…with about half of the school leaders in the survey citing the removal of SATs as the main reason they no longer paid as much attention to science'* (p.9). There is a resulting variation in the amount of time dedicated to the subject in different schools. In their 2015/16 Annual Report, Ofsted (2016b, p.47) found that the majority of schools spent between one and two hours per week on science; around a fifth of schools, however, spent less than an hour on this core subject. The pressure on primary schools is such that some even take the extreme step of focusing purely on English and mathematics in Year 6 until after the completion of the SATs. Overall, this will inevitably mean that some Year 7 pupils will have undertaken much more science in primary school than others.

The removal of the Key Stage 2 Science SAT also means that the assessment of primary pupils' attainment in the subject is undertaken solely by the school. Again, there will be variation as to how this is undertaken, the moderation procedures in place and how much detail is transferred to the secondary schools. The ever-changing landscape in English schools certainly makes teaching a challenging profession and the removal of National Curriculum levels from 2015, whilst having many advantages, has added a further complication into this equation. At the end of Key Stage 2, schools are only required to submit data centrally as to whether each pupil is working at the expected standard or has not met the expected standard in science. However, in order to monitor and demonstrate pupil progress, schools have adopted a plethora of different approaches to assessment. It is a spectrum, ranging from the use of computer-based tests to the assimilation of the range of formative assessment information collected throughout the pupils' time in school.

Whilst time is always at a premium, communication between primary and secondary is necessary, *'focused on the curriculum and underpinned by a shared understanding of the ways learning is assessed and the language of assessment'* (Ofsted, 2016b, p.50). This is essential to build up professional trust, as well as understanding, so that pupils' previous prior knowledge, understanding and skills are valued and built upon – not merely laid to one side so that they begin again in secondary.

It is difficult to overestimate the changes that Year 7 pupils experience when entering secondary school. They have been big fishes in a small pond and now they are the youngest – small fishes in a much larger pond. They are used to an environment where the norm is that they have one classroom teacher who knows them well, and now they have a range of teachers, some of whom will only meet them once a week. There are obviously many pastoral needs to be met but, in addition, there needs to be a clear focus on pupils' academic needs. One school expressed this succinctly: *'the secondary learning journey starts in Year 6'* (Ofsted, 2015, p.18).

There are many ways of easing this transition, such as:

➢ Referring to the Key Stage 2 Science National Curriculum when mapping Key Stage 3 so that the progression in knowledge and skills is made explicit;

➢ Cross-phase meetings:
  ◆ to build up professional trust and understanding
  ◆ to moderate assessment at Key Stages 2 and 3
  ◆ to work with feeder primary teachers when mapping Key Stage 3; and

> Cross-phase visits:
> ◆ Secondary science teachers visiting Year 6 primary science lessons to observe pedagogy and find out what subject knowledge and skills primary pupils possess
> ◆ Year 6 teachers visiting Year 7 lessons.

The development of cross-phase academy trusts opens up numerous interesting opportunities for some of the above. Realistically, though, time is a precious commodity and the key is to be a 'Trojan mouse'. The Trojan horse approach is to solve all the problems at once: laudable, but not always feasible. Being a Trojan mouse means scurrying around and doing small things that make a big difference. This might mean picking just one suggestion in the above list – the one that stands out as being a realistic starter.

Cross-phase observations will give real insight into the similarities and differences in pedagogy between primary and secondary. A key way of easing transition is to develop a continuity of the pedagogy, so that there is not a step change in the style of teaching that pupils experience. This could take the form of a short, dedicated discussion slot in science lessons, which gives the pupils opportunities to express their existing knowledge and understanding verbally – an excellent opportunity for elicitation, as well as stretching the pupils' thinking. A 'Bright Ideas Time' is something that many primary teachers include in their lessons (see https://pstt.org.uk/resources/cpd-units/bright-ideas-in-primary-science). This could involve, for example, the use of a concept cartoon, a 'big question' or a 'what if?' prompt.

Stepping into the science laboratory for the first time must be an exciting but potentially unnerving experience for Year 7 students: they will need to learn new rules in order to be safe, learn new ways of working, and they are about to don goggles and meet the Bunsen burner. Taking this as an example, having had the excitement of having learnt how to light and use it safely, it is often the case that they are then required to:

> Draw and label the Bunsen burner;
> Copy the safety rules into their books; and
> Learn the safety rules for a test.

Obviously, it is vitally important that they understand how to be safe, but taking a more 'primary' approach could be helpful by first asking the class:

> What safety precautions would you suggest? Why?

This will help them take ownership and potentially gain more understanding of why the rules exist. There is also the potential for a big question, such as:

- ➤ What is a flame?
- ➤ What happens to the gas? (We can smell it before it is lit but not after it has burnt).

It may feel counter-intuitive to ask such questions before the related teaching has taken place, but it allows for the pupils to be intrigued and to postulate. It will also reveal a range of existing understanding. They could do some research for homework or it could be left open, on the understanding that science is all about questioning and that the answer to these and many more interesting questions will be addressed during their secondary science experience.

Science is a great subject: it is all about thinking and questioning, something that we must surely encourage in our pupils at both primary and secondary levels. As Burdett and Weaving (2013, pps.2-3) state so well, *'we need to make sure that our science curriculum not only delivers students who "know facts" but also young people who understand and can use that knowledge…science education is not just about good performance in tests; it is about sparking interest in science at an early age and encouraging young people to continue to study science'*.

In practical sessions, allowing the pupils to make suggestions as to how investigations can be carried out and, sometimes, to plan and carry out their own, will help to elicit their existing skills and understanding, and to build on what has gone on before in primary school. The evidence is clear that doing science experiments is a great motivator for all pupils (Wilson & Mant, 2011).

In conclusion

So, the secondary learning journey does indeed begin in primary and the more primary and secondary can work together to smooth the transition and facilitate the continuity of that journey the better. We have a common aim, as Harlen (2015, p.7) expressed so well: *'Throughout the years of compulsory schooling, schools should, through their science education programmes, aim systematically to develop and sustain learners' curiosity about the world, enjoyment of scientific activity and understanding of how natural phenomena can be explained'*.

Further reading

Earle, S. & Serret, N. (Eds.) (2017) *ASE Guide to Primary Science Education (Fourth Edition)*. Hatfield: Association for Science Education

References

Burdett, N. & Weaving, H. (2013) *Science education – have we overlooked what we are good at?* (NFER Thinks: What the Evidence Tells Us). Slough: NFER

Harlen, W. (Ed.) (2015) *Working with Big Ideas of Science Education,* Trieste: Science Education Programme (SEP) of IAP

Lewis, A. & Smith, D. (1993) 'Defining Higher Order Thinking', *Theory into Practice,* **32**, (3), 131–137

Ofsted (2013) *Maintaining curiosity. A survey into science education in schools.* London: Ofsted

Ofsted (2015) *Key Stage 3: the wasted years?* London: Ofsted

Ofsted (2016a) *Specialist and non-specialist teaching in England: Extent and impact on pupil outcomes.* London: Ofsted

Ofsted (2016b) *The Annual Report of Her Majesty's Chief Inspector of Education, Children's Services and Skills 2015/16.* London: Ofsted

Rocard, M. (2007) *Science Education NOW: A renewed Pedagogy for the Future of Europe.* Brussels: European Commission. Retrieved from: http://ec.europa.eu/research/science-society/ document_library/pdf_06/report-rocard-onscience- education_en.pdf (accessed April 2017)

The Royal Society (2010) *State of the Nation report on 5–14 science and mathematics education.* London: The Royal Society

Wellcome Trust (2013) *The Deployment of Science and Maths Leaders in Primary Schools, A study for the Wellcome Trust.* London: Wellcome

Wilson, H., & Mant, J. (2011) 'What makes an exemplary teacher of science? The pupils' perspectives', *School Science Review,* **93**, (342), 121–125

SECTION 2:

Students: all learning science

CHAPTER 5

Educating students for the world: the role of emotions in learning science

Brian Matthews

Positively engaging students' emotions in science lessons can transform the teaching and learning experience for both students and teachers, and encourage students to continue to study science. This chapter explores how teachers can approach developing students' emotions in the classroom, and the subsequent benefits.

Background

We live in a fast-changing world in which technological change is a major driver. This makes it even more important that students can see science as exciting and fascinating and become interested in it. As a teacher, I wanted my students to experience the same awe and wonder that had attracted me to science. The science staff brought this to their teaching, where it contributed to girls and boys in my school opting for physics at roughly twice the national average.

Through introducing group work, collaborative learning and problem-solving, teachers are able to make the curriculum topics relevant to students in ways that encourage them to feel positive about science. These changes make science more personal and engage the children's emotions.

Teaching science for the future

There are many reasons for teaching science, from getting students to enjoy the wonder of the subject to understanding it. Students also need to be prepared for employment, but society is changing quickly, for the most part because of technological developments. There is a fragmentation of work, with a trend towards small businesses and more self-employment. This is called the 'Gig economy'. On the one hand it gives more freedom for people to select jobs and therefore to feel empowered (CBI, 2016). On the other hand, there is a worry that people could have no right to redundancy payments or to receive the national minimum wage, paid holiday or sickness pay, and so could feel insecure and a loss of power (Brinkley, 2016; TUC, 2016). There is also an increasing range of employment that can be made routine and hence automated, particularly in manufacturing and clerical fields. This automation often results in either low-paid, low-skilled jobs or fewer high-paid professional jobs. Another difficulty is that knowledge soon becomes out-of-date and, in many cases, we will be trying to educate students for future employment when we do not know what these jobs look like.

In order to enable students to be prepared for work, the Confederation of British Industry (CBI) has been working with employers to define the employability skills (CBI, 2009) and has produced a list of seven areas, the first four of which are:

1. Self-management – readiness to accept responsibility, flexibility, time management, readiness to improve own performance;
2. Team working – respecting others, co-operating, negotiating/persuading, contributing to discussions;
3. Business and customer awareness – basic understanding of the key drivers for business success and the need to provide customer satisfaction; and
4. Problem-solving – analysing facts and circumstances and applying creative thinking to develop appropriate solutions (CBI, 2010, p.24).

This list of skills indicates that, for students to be able to prosper, they need a set of social and emotional skills that enable them to be flexible, resilient and reflective. These are very similar to the skills predicted as being essential in the future, the *21st Century Skills* (Bellanca & Brandt, 2010; Beers, 2011), which are: *collaboration, critical thinking, creativity* and *communication*. Similarly, the European Commission adopted a new and comprehensive *Skills Agenda for Europe* (European Commission, 2016), which includes critical thinking, problem-solving and digital competences as key to allowing people to fulfil their potential as confident, active citizens.

For students to excel, in both employability and 21st Century skills, requires good social and emotional development, as well as a logical understanding of science. For example, you cannot co-operate with others unless you have social and emotional engagement. Problem-solving involves collaboration with others, often across diversity. Critical thinking involves developing the ability to accept criticism and suggestions from others without feeling affronted. Ideas and suggestions can come from people with a range of genders, sexuality and ethnic backgrounds, so students should also be educated to work with and understand equality issues. The above skills can all be used to help students key into feelings about science and so learn to enjoy science, as thinking skills and personal development overlap with emotional factors. Different terms are used by different authorities – *Wellbeing* in Scotland (http://www.gov.scot/Topics/People/Young-People/gettingitright/wellbeing), *Personal Development, Thinking Skills and Personal Capabilities* in Northern Ireland (http://www.nicurriculum.org.uk/curriculum_microsite/TSPC/what_are_tspc/framework/index.asp) and *Personal Learning and Thinking Skills* in England (QCA, 2007). These skills are all an integral part of science and of being a scientist. Clearly, in order to be able to achieve many of these, students must work in groups and their emotions have to be involved and developed. You cannot work effectively with others unless you can empathise and take criticism without getting agitated. A closer look at some of the points under these headings illustrates this further. For example, students must *'become more aware of and … articulate the links between their thinking, feeling and behaviour and how this impacts on the learning process'* (CCEA, 2009, p.41).

Hence the cognitive and emotional are intertwined. Most, if not all, scientific endeavour involves scientists working collaboratively and so requires these learning and thinking skills. These skills involve building individual and group relationships within our increasingly diverse culture within the UK. For example, males and females, and people from different social classes and ethnic and religious backgrounds increasingly interact in the scientific community. It is helpful to model this with student-inclusive groupings in science classes (Matthews & Sweeney, 1997) because so doing may also help counter the false stereotype of the scientist as an isolated genius. Collaborative group work has the potential of making science more popular, especially with girls, as it moves away from the impression that science lessons are remote, not personally to do with students themselves, and boring (Osborne & Collins, 2000).

In most cases, students require active learning (see Chapter 7) to be able to improve their personal learning, thinking skills and confidence. They also need collaborative learning, interacting with those around them, to improve their social and emotional literacy.

Science: both rational and emotional

In contrast to the call for students to develop socially and emotionally as well as logically, science is commonly presented to students as a rational activity. The National Curricula commonly reinforce this view. For example, the National Curriculum of England and Wales (DfE, 2013) includes a section called *Working Scientifically*, which places an emphasis on planning experiments to make observations, testing hypotheses, carrying out experiments appropriately, and applying the cycle of collecting, presenting and analysing data. Hence, many students can, from their experiences in school, form the view that science is objective, neutral and non-emotional, and that science develops because scientists exclude their emotions.

As such, it makes no mention of the networking, collaboration and relationships that scientists develop as a central part of scientific development (Longino, 1990; Latour, 1999). An enhanced picture of how science develops should include both the rational and the social aspects, as shown in Figure 1:

Figure 1: How do people 'do' science? Taken from Matthews (2015, p196)

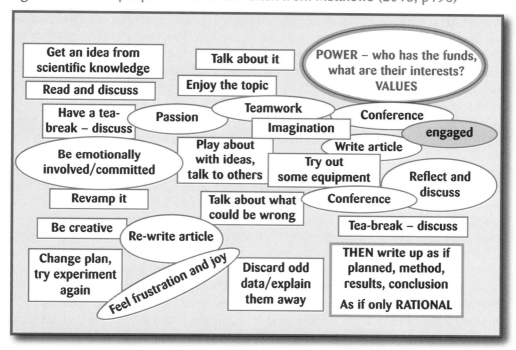

This chart illustrates the complex nature of the scientific endeavour (Matthews, 2015, p.196). Overlaid on this should be the slow uneven logical development of ideas that are, according to researchers who have watched scientists at work (Ziman, 1968), inseparable from the social and emotional changes taking place.

Zembylas (2005) argues that Western thought often embeds the idea that reason is opposed to emotion and that the detached and neutral knower offers the way to valid knowledge. Indeed, the whole school experience can lead students to believe that areas of schooling are rational and non-emotional. This reinforces in the students' minds the idea that learning is about rational cognitive development and therefore that they should not display 'feelings', especially in science lessons. Science teachers can work with students in ways that discourage emotional attachment and reinforce the impression that they should *not* bring emotions into science lessons – like repeating that science finds the facts – which are counter to what one would wish for in terms of students enjoying science. Teachers want students to like science and see science as a wonderful, interesting, imaginative and creative subject, yet these descriptive words *are* emotional and this means that students' emotions have to be engaged.

Teachers want students to experience a real emotional response to science, what it can do, and how it helps our lives. It involves sharing experiences so that students start on a journey of discovery – it is why students respond well to teachers who show enthusiasm and obviously love their subject. The teacher's display of emotions also helps to make science more personal to the student. These can become emotional habits (Trainor, 2008) where students are regularly taught to feel emotions in lessons so that it becomes a habit to learn with feeling.

The approach to 'working scientifically' is also out of step with philosophers such as Kuhn (1996), who have argued that science requires emotional commitment from scientists, otherwise there would be no progression in scientific understanding. This is because proof of an idea or theory requires a strong commitment. It is common for other scientists to be sceptical and question novel ideas; hence it would be easy for scientists with new ideas to abandon their theories unless they had a strong emotional devotion to them. Charles Darwin, for example, was very aware of the probability of strong criticism of his ideas from other scientists, as well as from society as a whole, and took decades to work on and clarify his theory of natural selection. It is therefore important that students engage emotionally with scientific ideas so that they are able to both understand and challenge them. We need strategies that allow students to accept science as an activity done by people, and to see it as both rational and emotional, with certainties and uncertainties, distinctness and ambiguity. If we can do this, then it will help in two ways. One, it will integrate the logical and social aspects of science and so help students to enjoy and relate to science more. Secondly, it can make science seem less remote and an activity that can be influenced. Hence students may take a greater interest in how science affects them and society. This is part of developing students' scientific literacy, or teaching *science for all* (Smith & Gunstone, 2007).

Some research evidence

There is evidence that engaging students' emotions in science improves their attitudes. For example, *Twenty First Century Science* engages students' feelings through discussing emotive issues that arise. Research into the impact of *Twenty First Century Science*:

'... provides striking evidence that GCSE courses which highlight the relevance of school science to out-of-school contexts and encourage discussion of socio-scientific issues, while also treating key theoretical ideas in the sciences in depth, increase student engagement and the numbers choosing to continue the study of science' (Millar, 2010 p.67).

Research has shown that students can enjoy science more and help each other to learn (Matthews, 2005; CASEL, 2008). Between 2004 and 2006, I carried out an intervention study in Year 7 (age 12) classrooms, which enabled students to develop their emotional literacy in science lessons. This helped the students to get on together, and understand more about their peers and what they thought. The students also helped one another to learn, so enabling them to mature (Matthews, 2005; Matthews & Snowden, 2007). Increasing emotional literacy in these classrooms resulted in students having an increased interest in science lessons and an expressed wish to continue with the subject. Tables 1 and 2 give the results of six parallel mixed-ability classes that were in research (82 students) and control (83 students) groups:

Table 1: Interest in science lessons (from Matthews, 2004 p.294)

	Percentage that are happy to see science on the timetable		Percentage saying science lessons are interesting		Percentage enjoying science lessons	
	Boys	Girls	Boys	Girls	Boys	Girls
Research	83%	85%	85%	88%	90%	88%
Control	73%	74%	77%	83%	76%	82%

Table 2: Continuing with science (from Matthews, 2004 p.295)

	Percentage that indicate they are likely to continue with science	
	Boys	Girls
Research	85%	85%
Control	71%	76%

On all factors, those students who had developed their emotional literacy were more positive about science and continuing with science. These results, along with those from *Twenty First Century Science*, indicate that engaging students in scientific issues for discussion and involving their emotions is likely to make them more interested in the subject.

Defining emotional literacy

Emotional literacy or intelligence is usually defined to include factors such as being able to know one's own emotions, manage these emotions, empathise and forge relationships (Steiner with Perry, 1997). Primarily, maturing is a process involving reflection, with different viewpoints being expressed; one needs other people to say what they thought happened and so challenge one to think about the discussion. Other people can give a different perspective, particularly if the person is of a different gender, ethnic background or social class. So, I define emotional literacy as follows:

'.... Emotional literacy involves connections between people and working with their differences and similarities while being able to handle ambiguity and contradiction. It is a dynamic process through which the individual develops emotionally and involves culture and empowerment. For example, it includes understanding how the nature of social class, "race" and gender (sexism and homophobia) impinge on people's emotional states to lead to an understanding of how society could change' (Matthews, 2005 p.178).

As a result, it includes equity issues as well as the usual factors, which is important for science teachers to realise as the take-up of science subjects can be overlaid with sexist stereotypes. This may seem difficult, but it can be easily achieved in the classroom.

The underlying principles for emotional development are that students can communicate with each other in a safe environment and think and reflect on the social processes and feelings that occur in collaborative learning. They should then:

- ➤ be able to verbalise (through writing and talking) what the interactions meant to them, and compare this with what other people thought had gone on;
- ➤ have some evidence about different feelings and can discuss their perceptions so that they come to understand their own and each other's viewpoints, both emotionally and cognitively;
- ➤ learn about each other and empathise with each other; and
- ➤ learn to understand the academic subject and become aware that knowledge generation and comprehension involves social and emotional interactions.

Classroom practice

One problem is that teaching cognitively is mixed in with emotional development, yet the two require quite different pedagogies. Provided that teachers are aware of it, this can be an opportunity to engage students with science and increase their interest in the subject. Students' emotions can be engaged, both in the science that they are learning, and between each other, through the use of collaborative group work (Matthews *et al*, 2002). The following example draws on the research and development work of the author and is described in detail in *Engaging Education* (Matthews, 2005).

Students are put, as far as possible, into mixed groups of 4-5. In co-educational schools, there should be a minimum of two of each gender. One student elects to be the observer, who reports back to the group on the interactions that happen in the activity.

The student-observer has a 'Discussion assessment' sheet upon which they record each time one of the group talks, and how much they interrupt, listen, support each other and what they learnt. They can also record anything else that went on.

The group has to carry out a collaborative task: say, complete a worksheet or a practical. After the activity, each student writes down on a separate sheet what they thought went on in the group, *without* talking to anyone. This is so that their perceptions of what went on are made explicit.

The student-observer then collects all the sheets from the students and runs a discussion based on what they and the others in the group have recorded. This legitimises each person's perceptions and a debate can ensue about learning and emotional and social interactions. It is through this process that students are able to develop empathy and understanding of how their actions make others feel.

The observer puts all the sheets together and gives them to the teacher to look at.

This process works well if it can take place about once every three weeks. The students usually take about half a term to get used to the procedure and eventually the student observer is removed, because the groups begin to work better together due to their improved emotional literacy.

Teachers can use evidence from the student sheets to focus on particular learning behaviours that can help the groups work more empathetically. Suppose a class has been working rather noisily. The teacher could ask them to get out a piece of paper and answer, by themselves, why they were working so noisily. Then, after they have been writing – in silence – but before they discuss their viewpoints, the teacher could suggest some options for the students to consider: 'Was it to do with the group?', 'Was it because of something that happened before the lesson?', 'Was it to do with the lesson itself?' and so on. This process helps students to reflect on their feelings towards each other and how well they get on. Students then have their feedback discussion and the teacher collects the replies to look over. In all cases, while the students are talking to each other, the teacher can go round and pick up evidence about how the students are getting on with both science learning and each other. At the same time, the students consider how much they have learned about the science and this legitimises an emotional engagement in science lessons.

In one example, students were asked about their attitudes to group work in science lessons – in this case, 'group work' meant their work with feedback discussions in which they talked about their attitudes. The results are shown in Table 3. Note that the boys' and girls' responses were similar in percentage terms, except for the first question.

Table 3: Students' attitudes to science lessons and group work

	Negative replies/%	Neutral replies/%	Positive replies/%
Does group work affect how you feel about science?	18% (boys 13%, girls 23%)	14% (boys 5%, girls 23%)	68% (boys 83%, girls 54%)
Would doing regular group work make students more interested in science?	11%	20%	69% (boys 71%, girls 67%)
Is group work a good way to learn science?	0%	11%	89% (boys 89%, girls 89%)

Here are some quotes from the students:

'Because you can compare answers and if you are right it will make you feel good' (girl)
'Because you can work with feelings' (boy)
'Group work is good, because if you don't understand the others will explain it and make it easier. Sometimes there are disagreements, but most of the time it is interesting' (girl)
'More interesting because you talk to other people' (boy)
'It makes it easier because you get ideas from other people. There is (sic) more disagreements because we all got different ideas' (boy)
'It is interesting, easier and more social' (girl)

The last quote is interesting as it reflects a general point that came out of the research – that girls found having the discussions made the lessons more social and so they connected with science more. This again suggests that student attitudes to science can be affected positively. One of the most significant findings was that the students helped each other to learn. Now, if boys and girls learn science together, and help each other to learn, then there is a greater chance that they will see each other as scientists – since they have learnt and worked at science together – and so both might be more likely to continue with science post-16.

The teachers on the project found that, in general, the classes became easier to manage and this carried over into the form groups. A teacher quoted: *'Certainly, from what I've seen pastorally, they emotionally support each other. When someone's having a bad day, others rally round, they are extremely caring, very concerned about the wellbeing even of people they don't normally go to or get on with. When one of the boys is upset, a lot of the boys are concerned, not just the girls, so it crosses gender. It's also the same with showing emotions to each other and getting support from each other'* (Morrison & Matthews, 2006 p.15).

In many ways, developing the students' emotional literacy is a whole school issue, but tackling it in science is essential if we want to encourage all students to be successful in science. There is a huge range of published materials and ideas that can be explored (see *Further Reading* and *References*).

Key points when preparing classroom strategies and materials

Clearly, to have a pedagogy for developing the emotions requires quite different modes of working from transmission teaching or active learning, although there are overlaps with the latter. To develop emotionally, students need to undergo experiences that allow them to reflect on and talk about their interactions and to compare their views about classroom events with those of others. Students need time to ponder, to hold a dialogue and to reflect with peers. It needs to be legitimised through classroom experience that students can talk about their emotions and feelings. This takes time and students need to engage with their feelings at their own pace, rather than be taught a set curriculum.

In general, students' emotional literacy should not be assessed by formal means, as this risks making some students feel worse about themselves and can damage their self-esteem, especially if they get the impression that they 'are not maturing fast enough'. However, it is probably useful for teachers to consider what the likely changes will be in the classroom environment when emotional literacy improves, and work towards that with their students.

Teachers can then use this to monitor the change at whole-class level without inhibiting individual learners.

In conclusion

In order to educate students for the fast-changing world in which we live, it is important that students are encouraged to develop not only logical skills, but also social and emotional ones. The incorporation of emotional literacy and thinking skills into schools is an important step in improving education for students, and can provide an exciting, challenging environment for science learning. It is a real opportunity for teachers to help learners to engage productively with science. Generally, most science teachers have not been trained to engage students with their emotions and the pedagogy can initially appear more challenging than it is. The rewards can be great – improved behaviour, enhanced interest and improved uptake in science. Addressing emotional literacy in science lessons can counter the implicit rule that some students believe – that science is not about feelings. When students are able to see that science lessons involve emotions, toward both the subject and each other, then they are released to be able to 'feel' science. Overwhelmingly, the students in the research stated that, although they found it difficult having to work with everyone, they thought it was very important that they did learn to get on with each other. This is part of what education is all about and it means that science is more likely to be seen positively by students.

Further reading

Alsop, S. (Ed.) (2005) *Beyond Cartesian Dualism: Encountering Affect in the Teaching and Learning of Science.* Dordrecht: Springer

Bellanca, J. & Brandt, R. (Eds.) (2010) *21st Century Skills. Rethinking How Students Learn.* Bloomington IN.: Solution Tree Press

Department for Education and Skills (2007) *Social and Emotional Aspects of Learning (SEAL) for Secondary Schools.* Ref: 00043-2007DWO-EN-04. London: HMSO. Available at: http://dera.ioe.ac.uk/6663/7/f988e14130 f80a7ad23f337aa5160669_Redacted.pdf

Matthews, B. (2005) *Engaging Education. Developing Emotional Literacy, Equity and Coeducation.* Buckingham: McGraw-Hill/Open University Press

Matthews, B. (2015) 'The Elephant in the Room: Emotional literacy/intelligence, science education and gender'. In *The Future in Learning Science: What's in it for the learner?*, Corrigan, D., Buntting, C., Dillon, J., Jones, A. & Gunstone, R. (Eds.), (Ch.10), pps.193–212. Dordrecht: Springer

References

Beers, S. (2011) *21st Century Skills: Preparing Students for THEIR Future.* Available at: http://cosee.umaine.edu/files/coseeos/21st_century_skills.pdf

Bellanca, J. & Brandt, R. (Eds.) (2010) *21st Century Skills. Rethinking How Students Learn.* Bloomington IN.: Solution Tree Press

Brinkley, I. (2016) *In search of the Gig Economy.* The Work Foundation. Available at: http://www.theworkfoundation.com/wp-content/uploads/2016/11/407_In-search-of-the-gig-economy_June2016.pdf

CASEL (2008) *Social and Emotional Learning (SEL) and Student Benefits: Implications for the Safe Schools/Healthy Students Core.* Collaborative for Academic, Social, and Emotional Learning (CASEL). Available at: https://safesupportivelearning.ed.gov/resources/social-and-emotional-learning-and-student-benefits-implications-safe-schoolhealthy

CBI (2009) *Future Fit. Preparing graduates for the world of work.* London: Confederation of British Industry. Available at: http://www.universitiesuk.ac.uk/policy-and-analysis/reports/Pages/future-fit-preparing-graduates-for-the-world-of-work.aspx

CBI (2010) *Ready to grow: business priorities for education and skills. Education and skills survey 2010.* London: Confederation of British Industry. Available at: https://www.educationandemployers.org/wp-content/uploads/2014/06/ready-to-grow-cbi.pdf

CBI (2016) *People and Partnership.* Confederation of British Industry. Available at: http://www.cbi.org.uk/cbi-prod/assets/File/pdf/ETS_report_proof_X.pdf

CCEA (2007) *Active Learning and Teaching Methods for Key Stage 3.* Belfast: PMB Publications. Available at: http://www.nicurriculum.org.uk/docs/key_stage_3/altm-ks3.pdf

Department for Education (2013) *Science programmes of study: key stage 3 National curriculum in England.* London: Department for Education (DFE-00185-2013). Available at: https://www.gov.uk/government/publications/national-curriculum-in-england-science-programmes-of-study/national-curriculum-in-england-science-programmes-of-study

European Commission (2016) *Competence frameworks: the European approach to teach and learn 21st century skills.* Available at: https://ec.europa.eu/jrc/en/news/competence-frameworks-european-approach-teach-and-learn-21st-century-skills

Kuhn, T. (1996) *The Structure of Scientific Revolutions 3rd Edition.* Chicago: University of Chicago Press

Latour, B. (1999) *Pandora's Hope.* Massachusetts: Harvard University Press

Longino, H. (1990) *Science as social knowledge: Values and objectivity in scientific inquiry.* Princeton, NJ: Princeton University Press

Matthews, B. (2004) 'Promoting emotional literacy, equity and interest in KS3 science lessons for 11–14 year olds; the "Improving Science and Emotional Development" project', *International Journal of Science Education,* **26,** (3), 281–308

Matthews, B. (2005) *Engaging Education. Developing Emotional Literacy, Equity and Coeducation.* Buckingham: McGraw-Hill/Open University Press

Matthews, B. & Snowden, E. (2007) 'Making science lessons more engaging, more popular and equitable through emotional literacy', *Science Education Review,* **6,** (3), 86:1–86:16

Matthews, B. & Sweeney, J. (1997) 'Collaboration in the science classroom to tackle racism and sexism', *Multicultural Teaching,* **15,** (3), 33–36

Matthews, B., Kilbey, T., Doneghan, C. & Harrison, S. (2002) 'Improving attitudes to science and citizenship through developing emotional literacy', *School Science Review,* **84,** (307), 103–114

Matthews, B. (2015) 'The Elephant in the Room: Emotional literacy/intelligence, science education and gender'. In *The Future in Learning Science: What's in it for the learner?,* Corrigan, D., Buntting, C., Dillon, J., Jones, A. & Gunstone, R. (Eds.), (Ch. 10), pps.193–212. Dordrecht: Springer

Millar, R. (2010) 'Increasing participation in science beyond GCSE: The impact of Twenty First Century Science', *School Science Review,* **91,** (337), 67–73

Morrison, L. & Matthews, B. (2006) 'How students can be helped to develop socially and emotionally in science lessons', *Pastoral Care in Education,* **24,** (1), 10–19

Osborne, J. & Collins, S. (2000) *Students' and Parents' Views of the School Science Curriculum.* London: Wellcome Trust/Kings College. Available at: https://www.kcl.ac.uk/sspp/departments/education/web-files2/news-files/ppt.pdf

Qualifications and Curriculum Authority (2007) *A framework of personal, learning and thinking skills.* Available at: http://webarchive.nationalarchives.gov.uk/20110215111658/http://curriculum.qcda.gov.uk/key-stages-3-and-4/skills/personal-learning-and-thinking-skills/index.aspx

Smith, D. & Gunstone, R. (2007) 'Science Curriculum in the Market Liberal Society of the Twenty-first Century: "Re-visioning" the Idea of Science for All', *Research in Science Education,* **39,** (1), 1–16

Steiner, C. with Perry, P. (1997) *Achieving Emotional Literacy.* London: Bloomsbury

Trainor, J.S. (2008) *Rethinking Racism: Emotion, Persuasion, and Literacy Education in an All-White High School.* Illinois: Southern Illinois University Press

TUC (2016) *Living on the Edge: The rise of job insecurity in modern Britain.* Trades Union Congress. Available at: https://www.tuc.org.uk/sites/default/files/Living%20on%20the%20Edge%202016.pdf

Zembylas, M. (2005) 'Emotions and science teaching: present research and future agendas'. In *Beyond Cartesian Dualism: Encountering Affect in the Teaching and Learning of Science,* Alsop, S. (Ed.). Contemporary Trends and Issues in Science Education Series, (Ch. 6), pps.123–132. Dordrecht: Springer

Ziman, J. (1968) *Public Knowledge. An essay concerning the social dimension of science.* Cambridge: Cambridge University Press

CHAPTER 6

Planning for motivation and engagement

Mark Hardman

The challenge of how to engage students in science lessons, and how to motivate them to try hard in their learning of science, has been faced by teachers for many years. In this chapter, the author talks about some of the reasons why students are less (or more) motivated and how, as science teachers, we can share our love of science effectively and help our students to be more confident at learning it.

Introduction

Science education allows students to understand the world around them, to develop the skills necessary to interrogate evidence presented to them in society, and opens doors to a huge range of careers and interests. Teachers sometimes find it difficult therefore to see why young people are not engaged in science lessons. Both research and experience suggest several reasons though: students may find science abstract and divorced from their own lives and goals, they may have got the impression that they are not good at science, and even those that do well in science can still see it as 'not for them'.

Seeing a class grow in confidence and become more engaged over time is one of the most rewarding experiences in teaching, but can only be achieved through actively planning for motivation and engagement, and builds on from the previous chapter, which considers students' emotions. This chapter offers a framework for planning towards the motivation of students in the short term, and their engagement with science in the medium term. Firstly, I give a brief overview of motivational theory and then develop this in relation to contemporary science education.

Motivational theory in education

Rewards and sanctions

Many schools still deploy systems of sanctions and rewards in order to manage the behaviour of classes. This can be seen as having derived from behaviourist theory, in which animals (including humans) may be conditioned by providing rewards for desired behaviour (e.g. Skinner, 1965). However, over the last few decades there has been increasing evidence that motivations that are 'extrinsic' (coming from outside the individual), are problematic. Students can quickly end up focusing on the rewards themselves, rather than what they are intended to promote: learning. This is particularly true where the focus of reward and sanction is behaviour for behaviour's sake; a well-behaved student who answers questions and does all their work could still see very little value in it. One way to mediate this is to link praise to learning more closely, so, for example, to praise students' use of evidence or correct scientific terminology in an answer, rather than praising them for their contribution alone. As Covington and Müeller (2001) argue, rewards still have a place in education if they can be deployed in a meaningful way to support the interests of students. This means thinking very carefully about what is being rewarded, though, and how students are likely to react if they do receive that reward, and if they do not. For example, might students see it as unjust if only confident students are being rewarded, rather than those engaged in learning?

Intrinsic motivation and interest
Focus on the 'intrinsic', internal motivations of students goes back to at least Dewey's (1913) study of interests and effort. This should not be understood as just about the whims or momentary excitement of students, though; Dewey recognised that interest is something that must be gained and then sustained within schooling.

Drawing on earlier cognitive theories, 'conceptual change' research in science education developed in the early 1980s and 'cognitive conflict' remains a key feature of these theories, suggesting that surprising or novel events stimulate learning (Posner, Strike, Hewson & Gertzog, 1982). Potvin and Hasni's (2014) systematic review of research into interest, motivation and attainment in science education supports the potential that science has here; a focus on inquiry, collaborative work and relating science to 'real life' all has a positive effect.

Yet, despite the evidence for all of these approaches within science education, in the English speaking world at least there is a well established decline in attitudes towards school science during secondary school (Barmby, Kind & Jones, 2008; Potvin & Hasni, 2014). This is despite no decline in agreement that science is important. Whilst factors outside of school obviously influence this (Gokpinar & Reiss, 2016), it appears that many of our schools, curricula, assessments and teaching practice are conspiring to turn students off science. As Brophy (2004, p.13) argues, teachers are not able to follow the intrinsic interests of students when curricula and assessment arrangements impose what is valued.

Goal theory

Given the interplay between these two sets of motivations: extrinsic and intrinsic, it is important to consider students' goals and how these are mediated within classrooms and schools.

The reasons that students give for their motivation can be broadly separated into mastery goals or performance goals. Mastery, or learning, goals are where a student is motivated to learn in order to improve their own understanding or skill. Performance goals are those associated with a particular measure or comparison, the most prominent examples in schooling being grades and competition. Overall, the weight of evidence suggests that students oriented towards mastery goals are more likely to succeed than those concerned primarily with their own performance (Kaplan, Middleton, Urdan & Midgley, 2002). These findings stem from research into cultural differences in how achievement is defined, how young people understand ability, and how they respond to difficulty and failure (which is currently being given much attention as developing a 'growth mindset').

Students hold multiple goals and respond to situations in a complex way; however, those who are motivated more by concern for their own performance are likely to focus on avoiding damage to their sense of self worth (Covington, 1992). Urdan *et al* (2002) explain the ways that this can manifest: a student may not put in the effort that they could, so their (poor) performance might be linked to effort and not their 'intelligence' (this is known as *self-handicapping*). They may simply avoid seeking help, allowing them to go unnoticed without progressing their learning. Alternatively, they may avoid novelty or challenge by doing work they already know that they are capable of. In some cases, they may just cheat, and this could be as simple as asking someone else the answer or waiting for the teacher to provide it. This learned helplessness moves students away from activities that promote their own learning.

Developing an expectancy-value model of motivation in science education

In bringing together considerations of rewards, interests and goals of students in school, several authors have drawn on expectancy-value theory (e.g. Brophy, 2004; Elliott, 2005).

'Theorists in this tradition argue that individuals' choice, persistence, and performance can be explained by their beliefs about how well they will do on the activity and the extent to which they value the activity' (Wigfield & Eccles, 2000 p.68).

Over the rest of this chapter, I will develop a framework for considering how secondary science teachers might develop the expectancy that students have of success, and the value that those students see in learning science. I consider these in both the short and medium term, as represented in Figure 1. This 'research-inspired' framework draws on motivational theory, but also a broader appreciation of contemporary practice in science education, which in turn draws on both research and experience of working with and in science departments over the last 13 years. Central to this framework is the view that, in classrooms, we are always developing both the short- and medium-term goals of students, and these are intertwined. Furthermore, we have to actively plan for motivation, and this cannot be separated from planning to develop the understanding and skills of students. As Keller (2010) argues, it is effort, knowledge and skill that together determine how satisfied a student is with the outcomes of a task.

Figure 1: Developing Expectancy-Value in the Short and Medium Term.

Short-term

Developing Expectancy of Success

Differentiation
• By outcome using hierarchical criteria
• By tasks and flexible use of groups

Developing mindset
• Teaching about brain development and learning
• Linking effort, skills, asking for help and outcomes

Collaboration
• Sharing responsibility for learning

Showing the value of science

Framing learning with authentic context

Deploying genuine enquiry
 (where the answer is not known)

Promoting transformative experiences
• Consider the original significance of key ideas
• Foster students taking learning into their lives
• Use metaphor and model to re-see ordinary
  phenomena through a scientific lens

Medium-term

Developing Expectancy-Value

Elicit and value the culturally-specific understandings of science; link these to curricula learning

Explicitly teach the nature of science
• Its strengths and limitations in answering life's questions
• The social processes and impact of doing science
• The collaborative, often international and diverse practice of contemporary science

Embed realistic career-based scenarios to highlight the roles which 'do' and 'use' science (not just 'being' a scientist)

Promote a scientific community within the classroom, which values listening, critique, evaluation of evidence and self-regulation

Use outreach, visits, clubs and role-models to show students the opportunities and pathways to using science

Developing expectancy in the short term

The expectancy that a student has of being able to succeed in a task is to do with what they perceive to be involved, as well as how far they might experience difficulty. In the short term, teachers may equip students to deal with difficulties when they arise, and differentiate activities to provide a challenge with which students feel capable of engaging. I will deal with each of these strategies in turn here.

Developing mindset and collaboration

How far a student will engage with a challenge is entwined with their view of their own capabilities. In reviewing recent research in this area, Yeager and Dweck (2012) show that teaching students that intellectual abilities are qualities that can be developed, rather than fixed, tends to lead to higher achievement, and can also lower adolescents' aggression and stress in the face of difficulties. They report on interventions such as teaching students about how the brain adapts as we learn (even in adulthood), which is of course consistent with the science curriculum. They also review how linking such messages about growth and learning to study skills provides better outcomes than developing study skills alone. Successful interventions conveyed the message that brain development involves effort, using effective strategies to learn and help from others. Such focused interventions are relatively simple to embed in science learning, but should be coupled with teachers trying to not convey a view of ability as fixed, which often means looking beyond the way that data and student groupings are commonly used in schools.

An amassing evidence base shows that setting and streaming students by attainment has no positive effect for most students, and actively disadvantages students who are classified as 'lower ability', yet this has so far done little to change its ubiquity in England (Francis et al, 2017). The issues with ability grouping include a lack of fluidity, applying different curricula, pedagogy and assessment to groups and the impact of the perceptions of both the teacher and the students themselves.

Whilst change at the national and even school level may be slow, science teachers should recognise that group work can be used effectively to support the expectancy that students have in the short term. Put simply, sharing the responsibility for a task means that any difficulty does not rest at the door of one student. This may be as simple as allowing students to discuss their ideas with people sitting near to them before giving a response to a teacher, but can also be true of more sophisticated group work. However, there is also a need to ensure that we are still assessing individuals within groups (Slavin, 1996), and providing individuals with appropriate challenge.

Differentiation

Individual students will perceive tasks differently, dependent upon their self-belief and interest. However, this is dynamic and depends upon the specifics of a task and situation. This means that student interest and self-belief can be developed if they are provided with appropriate challenge through effective differentiation of activities. However, as Winstanley (2010) argues, 'challenge' is difficult to define and often used loosely. The key is in ensuring that students are guided towards the next steps in a subject, but this involves teachers knowing what is intellectually demanding with a subject, and knowing their students. This is discussed in Chapter 9, but vital to this is development of the expectancy that students have of success in the short term. This involves effective differentiation to provide appropriate challenges, and developing the recognition that intelligence is dynamically linked to effort, and that the way students are grouped together allows them to overcome difficulty.

Showing value in the short term

Context and transformative experiences

Expectancy and value are highly interlinked and tend to develop together (Elliott, 2005); however, science presents myriad opportunities for developing intrinsic interest, showing utility and linking to the lives of students. Embedding relevant context into a curriculum has been shown to improve attitudes towards science and reduce gender difference in those attitudes, whilst not reducing the scientific understanding achieved (Bennett, Lubben & Hogarth, 2007). However, embedding 'real life' scenarios needs to be genuine, and involves knowing your class and what they are interested in. I once taught waves entirely through reference to guitar strings, but my attempts to teach momentum using a skateboard resulted in groans! Teachers may feel that considering context wastes time for learning, or they fall into the trap of presenting the scientific concepts first before students can 'apply' them. I believe that context must be there from the outset of a topic and must drive it. I worked with a teacher who spent a whole lesson with a class taking apart the remotes of games consoles to reveal and discuss the components inside. She then taught several lessons on resistance and capacitance (potentially abstract topics), with the full attention of the class.

Science is about exploring and explaining the world, and so scientific inquiry provides a way to link science to the world, but also provides a view of what science is for. Zimmerman and Glaser (2001) found that students reason differently about whether tap water is bad for plants and whether coffee grounds are good for plants: how familiar they are with these situations and whether they are testing a negative or positive claim seems to make a difference. Interest is stimulated by *genuine* questions about the world and this is an important part of a student valuing an activity.

More recently, Pugh *et al* (2010) have developed the notion of 'transformative experience', which links together conceptual change and the value that students see in learning about something, but also considers the capacity for that learning to shed light on their broader experiences. They use the examples of a young girl realising that her bike is made of the shapes she is learning about in school, or a student developing a lifelong interest in birds through learning about Darwin's finches. Such transformative experience requires someone to transfer their learning to new contexts and in so doing expands their perception of those contexts. Sadly, Pugh and his colleagues find that transformative experience is rare in science education, and that fostering engagement in class does not automatically mean that students will see the implications of their learning outside the classroom. Although research in this area is relatively new, they do propose some ideas for fostering such transformative experience (Pugh & Girod, 2007):

➢ Restore the original significance of concepts by helping students to appreciate their birth. For example, by exploring why Newton's ideas about motion are important, and how they were once powerful and potentially dangerous (e.g. to religious views at the time), students might look past Newton's ideas as 'classics' that cannot be questioned.

➢ Foster students taking their learning into their own lives, not just by connecting content with those lives, but also by passionately advocating how the learning might change their view. For example, evolution might be linked to how every animal is amazing in its capacity to thrive in a particular environment. The point is to craft the way that learning is presented so that students anticipate using it in their broader lives.

➢ Use metaphor and teach students to re-see ordinary phenomena from a new perspective. For example, by describing erosion as a battle between Earth's resistive features and those influences that destroy it, students are encouraged to see this in the rocks, paths and buildings around them.

These suggestions amount to developing the 'aesthetic' experience of seeing the world through scientific insight: like art, science has the potential to change the way we see things that are taken for granted. This capacity to develop transformative experiences has to be built into the way that we teach science, however. Students need to see both the passion of the teacher and also that the root of that passion lies in science as a way of seeing the world afresh.

In a study of 166 high school biology students in the USA, Pugh et al (2010) found that those who had mastery goals orientation (focusing on learning for its own sake), and those who considered science to be important to their own identity, were more likely to have transformative science experiences. In this section, we have looked at how, in the short term, using context and showing students how science can change their view of the world is likely to lead to great engagement. However, linking this to a student's identity and developing his/her medium-term goals is not something that can be achieved within a single lesson, and we shall turn to how teachers might forge these links now.

Developing expectancy-value in the medium term

Linking science to identity

A growing body of research is shedding light on the factors that influence medium-term engagement within science education, and making clear the inequalities in our society in relation to the 'science capital' that young people and their families have:

'Science capital refers to science-related qualifications, understanding, knowledge (about science and "how it works"), interest and social contacts (e.g. knowing someone who works in a science-related job)' (ASPIRES, 2013, p.3).

The background that students have in science is useful in explaining the seeming paradox that students value science, but do not want to become scientists. This may result from a mismatch between the expectations that students have of science and what they encounter in the classroom, as well as scientists being seen as 'other' or 'brainy', or science itself as being seen as 'masculine' (Archer *et al*, 2010). Whilst teachers cannot influence the family backgrounds of students, efforts are underway to link these insights into science identity with classroom practices. Nomikou, Archer and King (2017) suggest that teachers should elicit the culturally-specific understandings that students have of science, value different identities in the classroom and link these understandings and identities to scientific learning. For example, this might involve valuing the understandings of farming or food preparation that students and their families have, and linking this to curricula.

Understanding how people 'do' and 'use' science

Linking science to the identities of students can promote transformative experiences, but often needs a better understanding of the nature of science. Research on student understanding of nature of science dates back at least 20 years (Driver, Leach, Millar & Scott, 1996), but there are renewed calls for its necessity to be recognised in relation to teaching students from different cultural and religious backgrounds. Billingsley (2016) argues that science cannot answer all the questions that society faces and, in our teaching, we should acknowledge how scientific evidence provides answers to many questions, but does not preclude other ways of considering the world. Science is also often mistaken for a reductionist view and the subtleties of scientific argument missed by students. For example, students might equate the role of genetics in shaping personality with a commitment to genetic determinism. It is especially important to recognise the nature, merits and limitations of science when working with students who do not see science as integral to their identity, and to address the stereotypical view of science that students may pick up from broader society. Valuing and 'doing science' can be promoted without giving the impression that the study of science is about becoming a scientist, or rejecting other ways of seeing the world.

The ASPIRES (2013) study found that students and families are not aware of where science can lead, and that seeing scientists as 'brainy', white, male and middle-class impacts on their expectation of success in science-related careers. One approach to address this is to embed realistic and authentic careers-based scenarios into teaching, and a large European research project is currently investigating this (see www.multico-project.eu). Take the example of a town considering diverting a river due to flooding. This might require input from a hydrologist, meteorologist, civil engineer, the Environment Agency and many other professionals who work with science, but also farmers, environmentalists, councillors, business owners and other members of the community who would use science in understanding the impact of such a project. We noted above that transformative experiences can be promoted by showing how key scientific ideas came to be accepted. This of course risks framing science as the product of lone (white, wealthy) men and this heritage must be recognised. However, contemporary science is much more of a team effort, and looking behind any contemporary discovery will reveal the role of men and women, people from different backgrounds and international collaborations. Embedding an understanding of the roles and careers that involve science will promote students seeing the role of science in their futures; they may still do and use science, even if they don't see themselves becoming scientists. This enhances the 'utility value' of engaging with science.

Being part of a scientific community

The sense in which science involves a community of people can also be brought into the classrooms to foster a supportive learning community. Building upon our earlier discussion of developing mindset and sharing responsibility, teachers can support learning by being what Brophy (2004, p. 29) calls *'an authoritative manager and socializer of students'*. Students expect teachers to care about them as individuals and for activities to be interesting and have value, but they also expect clear classroom structures. Within such a space, students can make choices and feel competent and resilient, but also relate to one another and be mutually supportive. Whilst all classrooms should be supportive communities, science teachers might model this on the scientific community, which (at its best) involves collaboration, peer review, evaluation and creativity. This requires the explicit development of skills in listening, critique and evaluating evidence. It also involves promoting self-regulation, whereby students can control and manage their own motivation (Schunk & Zimmerman, 2008).

Developing a 'scientific community' within your classroom also affords opportunities to engage with professional scientists. This includes outreach activities from universities and other organisations, clubs, trips and visits, or even students themselves undertaking genuine research projects (see www.researchinschools.org for example). These have all been shown to have an impact on interest, motivation and attitudes towards science (Potvin & Hasni, 2014). In working with university outreach in schools, I have found that many teachers misunderstand the role of these experiences, and sometimes complain that the students do not learn much science from them. The point however is not (just) to teach students the science behind an exciting activity; it is to show them that they too could spend their time exploring interesting phenomena, often with important social consequences. Contact with role models and mentors from the scientific community also shows students that all sorts of people do science, and that there are many pathways and opportunities to do so.

In conclusion

Almost all science teachers come into the profession with a desire to share their love of science and its potential to have a positive impact on both students and the world at large. We sometimes need to be reminded that helping young people to succeed in their science exams is only part of that process. We must also actively develop the confidence of students and promote the personal value of science in opening doors and in answering questions about the world. We want them to see how science can benefit society, and fit with their future identities, whatever they choose to do in life.

Further reading

Covington, M.V. (1992) *Making the grade: a self-worth perspective on motivation and school reform*. Cambridge, New York: Cambridge University Press

Nomikou, E., Archer, L. & King, H. (2017) 'Building "Science Capital" in the classroom', *School Science Review*, **98,** (365), 118–124

Wentzel, K.R. & Brophy, J.E. (2014) *Motivating students to learn (4th Edition)*. New York: Routledge

References

Archer, L., DeWitt, J., Osborne, J., Dillon, J., Willis, B. & Wong, B. (2010) '"Doing" science versus "being" a scientist: Examining 10/11-year-old schoolchildren's constructions of science through the lens of identity', *Science Education*, **94,** (4), 617–639. https://doi.org/10.1002/sce.20399

ASPIRES (2013) *Young people's science and career aspirations, age 10–14.* Department of Education and Professional Studies, King's College London. Available at: https://www.kcl.ac.uk/sspp/departments/education/research/aspires/ASPIRES-final-report-December-2013.pdf

Barmby, P., Kind, P.M. & Jones, K. (2008) 'Examining changing attitudes in secondary school science', *International Journal of Science Education*, **30,** (8), 1075–1093. https://doi.org/10.1080/09500690701344966

Bennett, J., Lubben, F. & Hogarth, S. (2007) 'Bringing science to life: A synthesis of the research evidence on the effects of context-based and STS approaches to science teaching', *Science Education*, **91,** (3), 347–370. https://doi.org/10.1002/sce.20186

Billingsley, B. (2016) 'Ways to prepare future teachers to teach science in multicultural classrooms', *Cultural Studies of Science Education*, **11,** (2), 283–291. https://doi.org/10.1007/s11422-015-9701-9

Brophy, J.E. (2004) *Motivating students to learn (2nd Edition).* Mahwah, NJ: Lawrence Erlbaum Associates

Covington, M.V. (1992) *Making the grade: a self-worth perspective on motivation and school reform.* Cambridge, New York: Cambridge University Press

Covington, M.V. & Müeller, K.J. (2001) 'Intrinsic versus Extrinsic Motivation: An approach/avoidance reformulation', *Educational Psychology Review*, **13,** (2), 157–176. https://doi.org/10.1023/A:1009009219144

Driver, R., Leach, J., Millar, R. & Scott, P. (1996) *Young people's images of science.* Buckingham: Open University Press

Elliott, J. (Ed.) (2005) *Motivation, engagement and educational performance: international perspectives on the contexts for learning.* Basingstoke: Palgrave Macmillan

Francis, B., Archer, L., Hodgen, J., Pepper, D., Taylor, B. & Travers, M-C. (2017) 'Exploring the relative lack of impact of research on "ability grouping" in England: a discourse analytic account', *Cambridge Journal of Education*, **47,** (1), 1–17. https://doi.org/10.1080/0305764X.2015.1093095

Gokpinar, T. & Reiss, M. (2016) 'The role of outside-school factors in science education: a two-stage theoretical model linking Bourdieu and Sen, with a case study', *International Journal of Science Education*, **38,** (8), 1278–1303. https://doi.org/10.1080/09500693.2016.1188332

Kaplan, A., Middleton, M.J., Urdan, T. & Midgley, C. (2002) 'Achievement goals and goal structures'. In *Goals, goal structures, and patterns of adaptive learning*, Midgley, C., Anderman, E.M. & Anderman, L.H. (Eds.), (Ch. 2), pps. 21–54

Keller, J.M. (2010. *Motivational design for learning and performance: the ARCS model approach*. New York: Springer

Nomikou, E., Archer, L. & King, H. (2017) 'Building "Science Capital" in the classroom', *School Science Review*, **98**, (365), 118–124

Posner, G.J., Strike, K.A., Hewson, P.W. & Gertzog, W.A. (1982) 'Accommodation of a scientific conception: Towards a theory of conceptual change', *Science Education*, **66**, (2), 211–227

Potvin, P. & Hasni, A. (2014) 'Interest, motivation and attitude towards science and technology at K-12 levels: a systematic review of 12 years of educational research', *Studies in Science Education*, **50**. (1), 85–129. https://doi.org/10.1080/03057267.2014.881626

Pugh, K.J. & Girod, M. (2007) 'Science, Art, and Experience: Constructing a science pedagogy from Dewey's Aesthetics', *Journal of Science Teacher Education*, **18**, (1), 9–27. https://doi.org/10.1007/s10972-006-9029-0

Pugh, K.J., Linnenbrink-Garcia, L., Koskey, K.L.K., Stewart, V.C. & Manzey, C. (2010) 'Motivation, learning, and transformative experience: A study of deep engagement in science', *Science Education*, **94**, (1), 1–28. https://doi.org/10.1002/sce.20344

Schunk, D.H. & Zimmerman, B.J. (Eds.) (2008) *Motivation and self-regulated learning: theory, research, and applications*. New York: Lawrence Erlbaum Associates

Skinner, B F. (1965) *Science and human behavior* (First Free Press Paperback edition). New York, NY: The Free Press

Slavin, R.E. (1996) 'Research on Cooperative Learning and Achievement: What we know, what we need to know', *Contemporary Educational Psychology*, **21**, (1), 43–69. https://doi.org/10.1006/ceps.1996.0004

Urdan, T., Ryan, A.M., Anderman, E.M. & Gheen, M.H. (2002) 'Goals, Goal Structures, and Avoidance Behaviours'. In *Goals, goal structures, and patterns of adaptive learning*, Midgley, C., Anderman, E.M. & Anderman, L.H., (Ch. 3), pps. 55–83

Wigfield, A. & Eccles, J.S. (2000) 'Expectancy-Value Theory of Achievement Motivation', *Contemporary Educational Psychology*, **25**, (1), 68–81. https://doi.org/10.1006/ceps.1999.1015

Winstanley, C. (2010) *The ingredients of challenge*. Stoke on Trent: Trentham Books

Yeager, D.S. & Dweck, C.S. (2012) 'Mindsets that promote resilience: When students believe that personal characteristics can be developed', *Educational Psychologist*, **47,** (4), 302–314. https://doi.org/10.1080/00461520.2012.722805

Zimmerman, C. & Glaser, R. (2001) *Testing positive versus negative claims: A preliminary investigation of the role of cover story on the assessment of experimental design skills* (CSE Report 554). University of California, Los Angeles, National Center for Research on Evaluation, Standards, and Student Testing (CRESST). Available at: http://cresst.org/publications/cresst-publication-2926/

CHAPTER 7

Active learning

Ed Walsh

The two preceding chapters have focused on making lessons interesting and engaging for students; whilst these are both important, the learning process itself should also be active. But what does active learning look like? How can teachers plan and teach for active learning? This chapter addresses both of these questions and offers some concrete suggestions for teachers' practice.

Making learning active

It is a safe assertion that most teachers would declare themselves to be in favour of active learning, at least in principle. We would probably have to go quite a long way to find people in the profession who didn't see it as being desirable or who couldn't give examples of instances when they had succeeded in making this part of their teaching. However, that doesn't mean that it is universal or even common in science classrooms. Heavily loaded curricula and exacting and unimaginative assessment requirements are often quoted as factors that mitigate against learning being made more active. It is also true that developing the conditions for active learning needs skill and expertise. Like a Head Chef or the conductor of an orchestra, the 'leader of active learning' needs to do many things well and at just the right time.

Imagine going into a classroom or teaching laboratory part-way through a lesson. The lesson is well under way and the students are, at least at first glance, involved in some kind of activity. Would you know if the learning was active? Could we say, for example, that the learning was active if there was no practical work under way? Would there have to be discussion for it to be active? Would we say it was active learning if there were questions being answered in some way, even if it were individual work? Or would we be more likely to conclude that learning was active if there was movement, talk and an apparent sense of enjoyment? In order to form a clear view about whether active learning was a constituent feature of the lesson, we would need to find out a bit more.

Reflecting on this further, we might want to consider whether we regard active learning as being an essential feature of all teaching. Might we, realistically and with a pretty full teaching timetable and a lot of ground to cover, aim to 'deliver content' first, and then, time and energy levels permitting, supplement this with an active learning component – a kind of pedagogical booster rocket to add value? Or is it something that we can realistically aim for as being an intrinsic property of our teaching, present throughout? This chapter sets out to explore what we mean by active learning, why we might want to promote it and how we might go about developing its role in our lessons.

Understanding what active learning is

Let's consider an example. A teacher wants her students to appreciate that an argument to support the idea of continental drift is that land masses, now thousands of kilometres apart, can be made to slot together rather like a crude jigsaw. To make the learning more active, she decides, rather than just say this, possibly with the support of appropriate diagrams, to ask students to cut out the shapes of the land masses and try fitting them together. The students spend the next fifteen minutes cutting out shapes with various degrees of success and the following two shunting them towards each other. Was the learning made more active? There's a multi-faceted set of factors at work here. The teacher may have wanted to break up the more didactic parts of the lesson with something more kinaesthetic and the students might have appreciated the change of style of activity. It is difficult, however, to argue that the quarter of an hour spent trimming the shapes of continents increased the degree of understanding and therefore it is difficult to see it as a good example of active learning. Active learning is not simply 'doing activities whilst learning'.

It is important for us to consider what active learning is and what it isn't. It does not necessarily mean practical work, though this may be featured in lessons that have a strong component of active learning. Neither, as we've seen, can it be a euphemism for 'kinaesthetic interlude'. What it does refer to is the way in which students are engaging with activities. It represents a glittering prize – if achieved, it will embed understanding better than passive learning and it will develop and embed a broader range of skills. What characterises active learning is when the student *'rediscovers or reconstructs truth by means of external or internal mental action, consisting in experiment or independent reasoning'*. This isn't a new idea, incidentally; this quote is from Comenius, writing in the seventeenth century (quoted by Pinder, 1987, p.9). This leads us to a key point: active learning is not just more engaging or a way for a teacher to demonstrate a higher degree of professional competence, but is a more effective way of learning. Students who learn actively learn better.

Let's consider the idea of active learning more carefully. It clearly positions the student as the person who is doing the construction or the rediscovery – not the teacher who is showing the student these things. This isn't to say that the teacher doesn't have a crucial role in managing learning or that students are to be left without guidance to possibly stumble upon certain gems of wisdom. It also doesn't mean that the teacher *never* shows students explanations; rather, it means that the pedagogy has to be more than the teacher *always* showing students things. Science lessons shouldn't be like a coach tour in which everything, however clearly indicated, is always observed from behind the windows of the vehicle. Furthermore, the mental action referred to can, and should, be of a wide variety. It could relate to carrying out an experiment, participating in a discussion, responding to a news report or listening to a direct input from the teacher, or watching a demonstration. Whichever it is, the student has a central role in the sense-making by developing an explanation and, as they are making the links, they need to be exploring possibilities and seeing what works. These explanations might relate to theoretical or more practical contexts, but the explanation is developed by, rather than presented to, the learner.

The school science curriculum is sometimes criticised as 'giving students the answers to questions they haven't asked'. The idea of owning learning is central here – it is crucial and needs to be nurtured. For learning to be active students have to feel it is theirs. That's not to say that they selected the topic or even the activity, but that what they are then able to explain stems from their interpretation of the ideas. They have made the connection. This might entail selecting which evidence to focus upon, which words to use and which concepts are useful. Students suggesting why a knife cuts butter better if it is sharper and, if the same explanation helps, to suggest why it also cuts better if it is hot, may not have chosen the topic or the challenge, but their explanation is theirs, in a way that it wouldn't be if it was presented as a *fait accompli*.

It also needs to be appreciated that talk is key to active learning. If we accept that active learning involves students in testing ideas, then they need to have a mechanism for so doing. How can a tentative explanation be given a trial run? It will often be by students offering something verbally to one or more other people, seeing how it sounds and getting feedback. Talk is a powerful tool for students testing and developing their ideas, and this is discussed in length in Chapter 16. The classroom in which active learning has a strong role is one in which talk is present, effective and functional. In this, it is worthwhile investing time and effort; Michael and Modell (2003) assert that *'there is considerable evidence that, when learners are required to articulate their current understanding of the material (either to themselves, their peers or their instructors), learning is facilitated'* (p.16). This can't be achieved, however, by asking students simply to talk about what they think – the talk needs to be structured and we'll consider this later on in the chapter.

Understanding how learning might be passive

Active learning can be counterpoised with passive learning and what this might look like. It is perhaps more useful to think first about its characteristics and then its symptoms. Passive learning is learning (and it *is* learning) in which students are being exposed to ideas without engaging or interacting with them. It is learning without the learner making the connections and without anything of the student being brought to bear on the context: showing and accepting without owning. How would we recognise this in the classroom? There might still be practical activity and discussion; the effect of these (possibly irrespective of the stated objectives) would be different though and tailored towards adopting a given set of ideas. The difference is in terms of progression towards a certain kind of outcome, one that involves making connections and developing explanations. It therefore involves assessment, and we'll see later in this chapter how assessment has a fundamental role in enabling the teacher to decide how active the learning has been.

Therefore, the continental drift example involved less in the way of active learning than the knife and butter example, even though the former involved more physical activity. The effective cutting out of shapes might be considered successful if the teacher's objective was to break up the lesson, or to develop a specific set of fine motor skills, but less successful if the purpose was to understand how the land masses might have fitted together; cutting out South America is unlikely to engage students in a discussion about the likelihood of that being the case.

Active learning involves students testing models

Effective learning in science involves the testing of models, which means that active learning has a key function. It is important that, as teachers, we understand what is meant by models. A model is a representation of the physical world so, for example, the equation for calculating speed from distance and time and a diagram showing the water cycle are models, just as a physical model of a cell is a model. Scientists work with models a great deal and, as science teachers, we not only use scientific models, such as equations of motion, but also teaching models, such as representing current flow in a circuit by asking students to stand in a circle and make a loop of rope move through their hands. The models we use have a wide range of forms and therefore testing them can look and feel quite different, but they are nevertheless being tested. For example, the idea 'forces cause motion' is a simple model and one that we would want secondary students to refine. We might want them to get to the point of understanding that 'unbalanced forces cause acceleration' and 'balanced forces cause speed to be constant, possibly but not necessarily zero, and in a straight line'. Unlike 'forces cause motion', however, these are counterintuitive and careful teaching will be needed to get students to make this journey. It will need input from the teacher, ways of challenging their existing ideas and the opportunity to refine and improve their models. They need to be testing models and seeing if they work with a variety of contexts, both from primary and secondary evidence.

Passive learning might involve telling students about Newton's laws of motion and even asking them to memorise them, but this is unlikely to get students to reach for a better model when faced with an unfamiliar context. Active learning will have challenged students to see if their earlier model is good enough to explain a variety of phenomena, and encouraged them to develop and take ownership of better ones.

Science is a multi-faceted discipline and students learn not only different things, but also different *types* of things. They learn to compare and contrast, the difference between meiosis and mitosis, for example, how to construct a conclusion from practical evidence and practical skills. Michael and Modell (2003) suggest that *'in thinking about learning, it is important to recognise that there are at least three different kinds of "things" that can be learned: (1) declarative knowledge, (2) procedural knowledge and (3) psychomotor skills'* (p.5). They define declarative knowledge as *'the "what" of a particular topic'*, consisting of facts, data, concepts, principles and relationships. Procedural knowledge is the *'how'* to do certain things, and psychomotor skills *'the ability to do things in the physical world'* (p.5).

Because there is a range of kinds of 'things' that can be learned in science, we need to ensure that the concept of active learning applies to each and recognise how it is likely to be different according to each kind. Each kind of 'thing' can be learned through teachers supporting students in testing models and developing explanations and, in some (particularly authentic) contexts, different types of learning will be in use and often integrated, with one type of learning supporting the development of another. For example, the classroom in which psychomotor skills are developed actively but then used to support entirely 'recipe-driven' practical work is one in which the practice is only partly being embraced.

Scientific inquiry and active learning

As we would expect, active learning has a strong relationship with scientific inquiry in that this should involve students not only in answering questions, but also in *working out how* to answer questions. The questions themselves don't necessarily need to originate from students (though it is good if they sometimes do), but students should become more competent and confident in knowing how to go about answering a question. Students should learn that inquiry involves a range of cognitive processes and that these should be selected and used to successfully arrive at a solution.

Goldsworthy et al (2000) concluded that scientific inquiry should be seen as a cognitive process, using thinking strategies to conduct investigations successfully. We need to be careful, though, to ensure that practical work doesn't always have an exclusive focus upon process and skill. Students need to appreciate and experience how investigations lead to the development of concepts.

We should not assume though that, if students are involved in an activity that involves the use of apparatus, it is a *de facto* instance of active learning or, indeed, that it needs to be recast in this way. For example, the first time that students use burettes for carrying out a titration, it might be appropriate to walk them through the necessary steps in a very closed and didactic way. This is not wrong and it can be an effective way of coaching practical skills that they then draw upon in an inquiry-based activity; we just need to be careful about referring to it as active learning because students are manipulating apparatus. This is also problematic when practical activities are completed as recipe-driven sequences. It should be obvious from our earlier consideration of the nature of active learning that, if no explanations are being developed and no models are being tested *in the minds of the students,* it is difficult to argue that the learning is active. Millar (2010) asserts that it is essential that practical work should not only be 'hands on' but also 'minds on' and this resonates strongly with the development of active learning (see Chapter 12 for further consideration of effective practical work). What effective teachers will often do is to make skilful use of questions, engaging students whilst activities are being carried out. Good practice here is to use both closed questions (e.g. 'why does the thermometer need to be positioned there?') and more open questions (e.g. 'why might this procedure be useful in determining water purity?').

Similarly, active learning can be non-practical activities. For example, if we pose the question 'why can it be a challenge for astronauts to maintain their muscle tone whilst working in microgravity?', it could involve a significant degree of active learning in terms of research, discussion and presentation of ideas, without any equipment in sight. If students go away with the idea that learning is only active if there are test tubes around, then their curriculum is a restricted one.

Climate for learning

Setting the climate for learning in the promotion and deployment of active learning is crucially important. Promoting active learning isn't only a case of designing or adopting certain types of classroom activity (though that is important), but also setting up the classroom with protocols and expectations that will mitigate towards learning being active. It can certainly be argued that the transition from passive to active learning is more difficult than sustaining an ethos of active learning once established.

These conditions need to be recognised and established; the effective teacher knows what they are and how to manage them. They include creating an environment in which tentative ideas can be shared without fear of ridicule, in which students feel that their own ideas and questions are valued and respected and in which dialogic talk has an established role. This is more than just establishing ground rules for discussion and debate, though those steps may well be helpful; it is also about establishing expectations about how ideas are explored. For example, when students have carried out an experiment and gathered data, they should discuss what conclusions can be drawn and how confident they can be. Challenging each other is part of this and is more powerful than everyone agreeing with an initially assertive voice. Students won't necessarily know this, though, so it will need to be explained and modelled.

It needs to be understood that science is about using processes and contexts as tools to develop explanations, rather than 'just learning stuff'. Again, there is much that the teacher can do to establish this, challenging students to apply ideas to a range of contexts. Care needs to be applied here, though. Sometimes students may not be used to being challenged in science or have learned that success (as in teacher recognition and test performance) comes from absorbing ideas rather than engaging with them. 'Challenging students' isn't a euphemism for 'picking arguments' however and the classroom should be a place where students feel comfortable, as well as one where active thought is valued. For more on this, see Chapters 5 and 6 on enjoyment and on motivation.

The role of outcomes

The relationship between the learning activity and the outcome is of paramount importance. We referred earlier in this chapter to the idea that including kinaesthetic activities in the classroom does not automatically mean that learning is active. It is important, of course, to be able to suggest how we can decide whether the learning we are facilitating is active. There are various ways of encouraging students to be active and various ways in which they can be active, but a central point is that the activity needs to be a plausible way of students being able to achieve the outcomes set.

To promote active learning, we also need to consider the type of outcome we are working towards. The design of the overall lesson or sequence of lessons will work better if outcomes are framed as behaviours – and if these are based on the skills and ideas that students need to succeed in their developing mastery of the discipline. If we think in terms of what students should be able to do, whether it's to focus a microscope, compose a testable question or apply their understanding of fields, then it's easier to ask if the learning activity will support the development of this activity. In many cases, the learning will be more effective if it's active because this will involve students in constructing explanations and testing models, which is at the heart of good science.

The role of assessment

The truth is, of course, that, however skilful we are at planning students' learning, we cannot know in advance how effective it will be. There are initial indications, such as whether students are engaged in the activity and whether they need to repeatedly ask for clarification. However, this won't tell us much; what we're really interested in is whether students are learning. Waiting until some form of periodic assessment takes place is not a great idea; we need a much shorter feedback loop than that, so that we can fine-tune the lesson. Of course, this leads into the area of formative assessment and Chapter 18 will give a much fuller guide. What we should emphasise here, though, are two prime functions of assessment in relation to active learning.

The first is the importance of assessing student degree of understanding and misconceptions, to know where the learning focus needs to be to support them in making progress towards the outcomes. Earlier in this chapter we considered an example from the topic of forces and motion. It might be a good idea to preface work on this with a short activity to indicate what students think about the causal effects of forces on motion. One way of doing this would be to present students with simple drawings of straightforward contexts, such as a person pushing a crate across the ground at a steady speed and asking students to suggest what forces are acting, in which direction they are acting and how they compare in size.

The second is to assess progress to work out if the active learning was effective. This isn't a cue for excessive formalised testing; indeed, effective teachers have understood for a long time that astutely-chosen activities can not only promote learning but also simultaneously reveal progress and provide assessment evidence. For example, students explaining whether they would use a series or a parallel circuit in a particular application, and setting up equipment to demonstrate this, would show if they had mastered key differences between these two types of circuit and understood their uses.

Dialogic teaching also has a crucial role here so that students' ideas are intentionally brought to the fore. To be effective the dialogue needs to be used proactively by the teacher and with a sharp focus on the key ideas and processes. What question(s) will 'cut to the chase' and indicate whether students have mastered a key idea?

Planning for active learning

Finally, it needs to be understood that active learning should be planned for. If it is going to happen, it will usually be that a teacher has planned and taught a curriculum that both establishes the conditions that apply generally to their classroom practice and also plans for explicit activities that support active learning. This is more likely to be effective if it is part of the ethos and practice of the subject team (and, ideally, has some resonance with the whole school). Students' expectations of a subject are directly influenced by their prior experiences so, if the science department is good at developing these, it will support active learning in all lessons. It will also mean that lesson plans and resources are more likely to be ones that directly support an active learning approach, thus reducing the onus upon an individual teacher to rework everything that is common to the department. Also of significance to a less experienced teacher, it will also be that, if and when issues are encountered, it will be easier to refer to colleagues.

In conclusion

It is hoped that a strong case has been made for active learning and a useful explanation provided as to what it is and how to go about setting it up. To go back to our original question about recognising active learning, when stepping into a colleague's classroom, it should be clear that some indications are more insightful than others. We couldn't really tell simply by seeing if students were involved in a kinaesthetic activity or if they were involved in discussions. We might need to linger in the room for a little while and listen for students developing their own explanations, making connections and putting forward their own ideas and interpretations. This could involve practical activity, discussion, research or other approaches – but they are actively 'trying out some ideas and seeing if they fit'. Students would feel able to construct solutions and refine them, whether it related to practical skills, the processes of science or specific concepts.

What also helps is a clear sense of the outcomes being aimed for. We might get a sense of that from what the teacher has displayed, but we know that it is really working well when the students themselves can tell us what they are doing and working towards. There would also be a sense of the teacher 'keeping their finger on the pulse'.

Active learning is a high level teaching approach that needs mastery of a range of skills and an astute sense of their deployment. It is advocated, though, because it works.

Further reading

ASE series on Professional Issues *P4.2 Active Learning*. Available at: https://www.ase.org.uk/resources/scitutors/professional-issues/p42-active-learning/

Active Teaching and Learning Approaches in Science (ATLAS). Available at: https://www.stem.org.uk/elibrary/collection/3635

Waldrop, M.M. (2015) 'Why we are teaching science wrong, and how to make it right', *Nature*, **523**, (7560), 272–274. Available at: http://www.nature.com/news/why-we-are-teaching-science-wrong-and-how-to-make-it-right-1.17963

References

Goldsworthy, A., Watson, R. & Wood-Robinson, V. (2000) *Investigations: Developing Understanding.* Hatfield: Association for Science Education

Michael, J.A. & Modell, H.I. (2003) *Active Learning in Secondary and College Science Classrooms: A Working Model of Helping the Learner to Learn.* Mahwah, NJ: Lawrence Erlbaum Associates, Inc.

Millar, R. (2010) 'Practical Work'. In *Good Practice in Science Teaching: What research has to say 2nd Edition,* Osborne, J. & Dillon, J. (Eds.), (Ch. 6), pps. 108–134. London: McGraw-Hill

Pinder, R. (1987) *Why don't teachers teach like they used to?* London: Hilary Shipman

CHAPTER 8

Planning for student understanding and behaviour for learning

Judith Hillier

Planning lessons is a key part of a teacher's role, and is crucial both for students' learning and behaviour. This chapter uses two theoretical frameworks to explore why planning matters and what it should involve, making some practical suggestions that should support the development of strategic and focused lesson planning habits.

Introduction

Lesson planning occupies a large part of a teacher's time: indeed, when I ask PGCE applicants what they think teachers spend their time doing when they are not in lessons, the first answer is always 'planning'. Many of us reading this chapter may already feel that we know quite a bit about planning and would rather spend time considering other parts of our professional practice. However, the process of planning lessons is both highly skilled and complex: taking the time to recognise and explore this may help us to identify a part of our practice that we would like to develop further, and will undoubtedly help those of us who work with and support beginning teachers who are learning how to plan. The frameworks provided in this chapter have been found by many teachers and teacher educators to be a powerful way of engaging with this skill, which is so fundamental to our profession.

Start with the child

As a teacher educator, I am regularly asked what my philosophy of education is: what do I think is the *best* way to teach? What is the optimum way to deliver content? Both of these are questions with which I am deeply uncomfortable. We have all seen strategies that have been used successfully by some teachers and less than effectively by others, fantastic resources that a teacher has found difficult to use and, conversely, a dull task brought alive by an inspirational teacher.

Similarly, the notion of 'delivering' lessons is one that I would reject completely: students are not empty vessels waiting for a teacher to decant knowledge into them; neither do we stand at the front and present information. At the very least, when we teach, we interact with students and encourage them to talk with one another and so enable them to engage in this social process of learning. However, as this whole book indicates, teaching is much more than that, and the process starts with the child: the student or group of students with whom you will be working and the funds of knowledge that they bring to the classroom. As described by both Wynne Harlen in Chapter 1 and Helen Wilson in Chapter 4, as teachers we need to think about prior science learning experiences that our students may have had in school and from their everyday lives, and the ideas about science that they are likely to bring to our lessons. We also need to think about who our students are as people: as discussed by Brian Matthews in Chapter 5, we should be aware of what they might enjoy, of their attitudes and feelings towards science and how that might impact on the way in which our students engage with our lessons. Both Chapters 9 and 10 highlight the diversity amongst our students – we all teach students who are diverse in terms of gender, class, ethnicity and SEND (Special Educational Needs and Disabilities), and this should be both celebrated and recognised as having a central influence on our planning. As teachers, it can take some time to build up this in-depth knowledge of our students, though experienced teachers will already have some understanding as a starting point. Beginning teachers need support with this: advice from the teachers of the classes with which they will be working, and time to spend with the students and observe them in lessons. All teachers would benefit from good school data systems, which enable the compilation of detailed and clear student profiles (see Chapter 19).

Use your professional knowledge as a science teacher

Having considered who our students are, the next question is what do we want to help them learn? For many of us, it can feel that this is dictated to us by the National Curriculum or specification statements from awarding bodies, although hopefully these will have been reworked by the school's schemes of work into medium-term plans (see Chapter 23). We might want to help our students to develop their understanding of current in series and parallel circuits, we might be asked to help them learn about elements and the Periodic Table, or about food chains and webs, or about how to plan and conduct an investigation into the reactivity of metals.

Whatever the learning objectives are, we now have our two starting points: *who are our learners* and *what do we want them to learn*. From these two points, all our other decisions follow, using our professional knowledge as science teachers about how to teach this idea to these students. The professional knowledge on which we draw is wide-ranging: we think about what it means to learn (Chapter 3) and how to make that learning active, rather than passive (Chapter 7); we think about to motivate and engage our learners (Chapter 6), including being creative in our teaching (Chapter 11); we think about how practical work (Chapter 12), ICT (Chapter 15) and other learning opportunities outside the classroom (Chapter 13) can be used to support students' learning; we think about the mathematical and literacy demands of the topic (Chapters 14 and 17), and the role that language and talk play in learning science (Chapter 16). We will also be influenced by health and safety considerations (Chapter 24, though hopefully not restricted by them!) and by our beliefs about the nature of the subject we are teaching (Chapter 2). Finally, our planning decisions will also include the process of using prior attainment data to inform our planning and assessing our students in order to plan how we might help them to understand what quality looks like within specific classroom tasks, and then use the evidence to monitor their progress and to inform our subsequent teaching (Chapters 18 and 19).

Hopefully, this long list will highlight the breadth of professional knowledge that teachers have and use every day when planning lessons – this is part of the intellectual endeavour of teaching, and is not to be downplayed. However, this is something of a clumsy way of thinking about it, and the next section will introduce two theoretical frameworks that a number of teachers and teacher educators have found to be helpful ways of thinking about our professional knowledge.

Theoretical framework 1 – Pedagogical content knowledge (PCK)

The first of these was proposed by Lee Shulman (1986, 1987), and was the first to explicitly recognise that teachers have highly specialised knowledge that is unique to their profession. Figure 1 (overleaf) shows the various categories identified by Shulman. A number of these will be familiar: as teachers, we know both our learners and our educational context – the school in which we teach and the community our school serves – this is shown on the left-hand side of Figure 1. We have an understanding of the purpose of education, and the role that it plays in our wider society. We will have developed general pedagogical knowledge about how to give clear and concise instructions, how to move students in and out of the classroom, how to gain and maintain students' attention – these all form the right-hand side of Figure 1. The middle section is the subject-specific part: our knowledge of the subject(s) that we teach is termed 'content knowledge'; curriculum knowledge is not just what we know about the curriculum and specification statements, but also our knowledge about how ideas and understanding are built up and developed as students progress through school. One way of describing this progression is given in Harlen's *Working with Big Ideas of Science Education* (2015, pps.22–33).

The final component of teacher knowledge is pedagogical content knowledge (PCK) or *knowing how to teach this particular content*. Shulman proposed that, when teachers know how to teach a certain topic, then they know the following things about this topic:

> *'what makes the learning…easy or difficult'*, including what ideas and understanding students are likely to bring to lessons – an excellent starting point with this is the summary of research into children's ideas, which has been reprinted over 20 times (Driver *et al*, 1994);

> *'the strategies most likely to be fruitful in reorganising the understanding of learners'*, which would include relevant practical activities for science teachers; and

> *'the most useful forms of representation…the most powerful analogies, illustrations, examples, explanations and demonstrations'* (Shulman, 1986 pps.9–10).

From this, it can be seen that Shulman subscribed to the constructivist theory of learning, described by Wynne Harlen in Chapter 1.

Figure 1: Categories of teacher knowledge proposed by Shulman (1987)

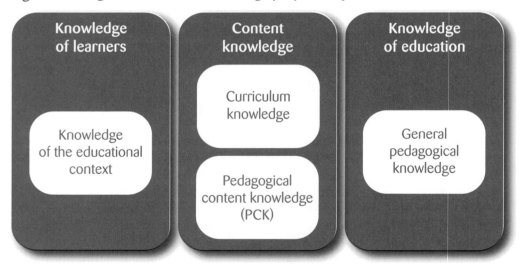

Much research has been done into PCK since Shulman suggested it (see Kind's 2009 review for a summary), with an ensuing debate about what it does and does not include. As a concept, some teachers have found it hard to relate to as, typically, it is explained using terms that seem like jargon and is rather an academic construct (Loughran, Berry & Mulhall, 2006). One helpful development of Shulman's ideas came from a project researching the teaching of electricity in primary schools (Summers, Kruger & Mant, 1998), which expanded the latter two points as follows (p.170):

Effective strategies:
- *'appropriate scientific terms and language to use with children'*
- *'simple technical knowledge of equipment'*

Useful forms of representations:
- *'what to emphasise'* in an explanation, both what is and isn't true
- *'how to simplify validly what are often very complex ideas'*

What does this mean for teachers when they are planning lessons? What I have described above is knowledge that many experienced teachers would recognise as key components of a 'scheme of work', or a 'scheme of learning' – the organised plans and resources that their science department has developed, or bought and adapted (see Chapter 23). As a beginning science teacher, your mentors and tutors will work with you as you develop your planning skills, and these questions should form part of the collaborative planning conversations that you have with those experienced teachers:

➤ What are the common misconceptions, preconceptions and alternative conceptions in this topic?

➤ What are the key learning activities, including practicals, that I should consider, and what do I need to know to ensure that the practical 'works'?

➤ What is the language of this topic, including the key words and phrases, which I should be using?

➤ What are some of the appropriate analogies, and what points should I emphasise in these analogies?

➤ Is my explanation of this concept both accessible for these students and still scientifically valid?

➤ How can I create an environment in my classroom that will encourage learners to think and make connections and possibly challenge some of the ideas?

Theoretical framework 2 – the Knowledge Quartet

As can be seen from the questions above, Shulman's model of teacher knowledge and the concept of PCK are highly relevant to planning and teaching, but no model can fully encapsulate all the complexity of teaching. Another model of teacher knowledge that has been found to be helpful is the

Knowledge Quartet, which was developed through research into the teaching of mathematics in primary schools. Analysis of videos and lesson observations of beginning teachers led to this framework, which has since been used to support teacher development (Rowland *et al*, 2009).

Figure 2 shows the four units of the Knowledge Quartet, with each unit comprising the following broad areas:

➢ Foundation – this includes scientific subject knowledge, beliefs about science and the purpose of learning science, and beliefs about science teaching;

➢ Transformation – this strongly references Shulman's notion about the *'capacity of a teacher to transform the[ir] content knowledge into forms that are pedagogically powerful* (1987, p.15) and includes knowledge about examples and analogies and ways of representing ideas that will be helpful to learners, with clear links to the discussion of PCK above;

➢ Connection – this includes knowledge about how to develop ideas within and across a series of lessons in a coherent way, about how to link to prior learning and to signpost future learning, creating a storyline to support learning; and

➢ Contingency – this includes knowing how to respond to children's ideas and questions, making use of opportunities and being able to deviate from the lesson plan.

Figure 2: The Knowledge Quartet, developed by Rowland *et al* (2009)

Transformation	Connection	Contingency
Foundation		

Again, experienced teachers would recognise much of the Knowledge Quartet as *tacit* knowledge that is crucial in their professional practice, but may not be discussed much within a school science department. As a beginning teacher, it is important to consider these issues: your beliefs about what science is and how we should teach will heavily influence your practice in the classroom (*Foundation*). Much of your time in your first years will be spent on transforming your subject knowledge into knowing how to teach your subject – the teachers around you will be hugely valuable in this, as will the ASE's *Teaching Secondary* books for biology, chemistry and physics (see Further reading) (*Transformation*). Learning how to sequence a lesson so that students' understanding gradually develops as ideas are built up into an explanation is one of the most important things you will learn as a beginning teacher, and should be one of the principles underpinning all your lesson planning (*Connection*). Planning for the unknowns is one of the hardest aspects of teaching, for obvious reasons, but this is also part of your teacher knowledge that you can develop. As a beginning teacher, there will be much that you won't expect and that you will have to deal with on the spot. Some situations can be anticipated, and experienced teachers will advise you of these; e.g. giving a class a set of metre rules will result in play-sword-fighting, no matter whether the students are 12 or 18, unless you set some clear expectations first! But there will also be genuine moments of contingency, when a student asks a question or gives a response that reveals a gap in their understanding, and you have to think on your feet in order to respond. Developing habits of dialogic questioning and classroom discussion (see Chapter 16) will help to foster an atmosphere of openness and will enable you to learn how to respond appropriately.

Pitch, pace and scaffolding

So far, the focus of this chapter has been on knowing how to plan lessons to support the development of students' understanding. But planning is also key to managing students' behaviour. Behaviour management is typically what beginning teachers are most worried about, and rightly so – the thought of how you, as a lone adult, can get 30 teenagers to listen to you and follow your instructions can be nerve-wracking! But teachers do this all the time and, every year, beginning teachers learn how to do this.

However, what sort of behaviour or behaviours do we want from our students? The notion of *behaviour for learning*, or behaviour that will be conducive to learning, has been much discussed (see Powell & Tod, 2004, for an early review). One way of thinking about it is to ask what is the 'normal desirable state' that I would like to have in my classroom? A normal desirable state is what we would like students to be doing *and* feeling for a particular part of the lesson in order to facilitate student learning. (See Cooper and McIntyre, 1996, for more detail.) Clearly, as discussed by Mark Hardman in Chapter 6, planning for motivation and engagement is crucial and Mark gives a clear picture of the need to plan so that students *expect* to succeed and *value* learning science. There are three aspects of lesson planning that are vital for this:

➤ Pitch – rather like Goldilocks, your lessons should not be too hard or too easy for the students in your class. Tasks need to be accessible, so students can understand what they need to do, and with enough challenge to keep them interested and learning, but not dispirited. Similarly, explanations need to be pitched at a level that the students understand, thinking about the language and scientific vocabulary that might need to be explained, how to simplify the ideas, and which analogies will students understand and find helpful;

➤ Pace – again, not too fast and not too slow. If you have planned your lesson with a clear storyline, it should be relatively easy to move from one teaching and learning activity to the next. For each one, you need to consider what is the key idea that I want all the students to know and understand, or skill they need to demonstrate, and how will I know they have done this? Once achieved, this is the point at which you should move on but, obviously, some students will reach that point faster than others, so make sure that you have a task or some questions that will get these students to develop their knowledge of this key idea further. Teachers often use Bloom's Taxonomy as a framework for this higher-order thinking (Krathwohl, 2002, and see https://www.tes.com/teaching-resource/bloom-s-taxonomy-teacher-planning-kit-6141531); and

➤ Scaffolding – this relates to both pitch and pace, and can be thought of as the support you as the teacher provide to the students through your examples and explanations, the ways in which you break ideas or questions down to help students understand, and the ways in which you model to students how to answer a question or do a calculation. Sometimes, these scaffolds will be given to the whole class as part of the pitching of the lesson. Other times, these scaffolds will be used to help support the students who lack some of the skills or prior knowledge of other students in order to help maintain the pace of the lesson. It is key here to remember why we are scaffolding; it is providing the support so that students can think and work out the ideas to which we are introducing them. Eventually, just as with any building works, you can remove the scaffold and their understanding will remain secure. Scaffolding is not about giving students the right answer to a likely question such as 'what do we need to do in an investigation?' where they learn to chant 'fair test', but cannot actually control and manipulate variables.

There are clear links between all three of these: pitch, pace and scaffolding, and the two models of teacher knowledge discussed earlier.

Building rapport

The final aspect of planning to be discussed is the question of how to build rapport with our students. In many ways, this will and should develop naturally, but this can also be supported by our planning. Students are more likely to trust and respect you if you look and sound as if you know what you are doing! Being sure of your subject knowledge, being clear in your use of analogies, having a coherent storyline to your lesson, and knowing how what you are teaching links to what they have learned before all help you to build rapport with a class. In the same way, using examples that are relevant and interesting to students (see Chapter 15) and to their lives both within school and within their community, taking account of students' emotional needs (Chapter 5) and developing an inclusive pedagogy (Chapter 9), will also help to build rapport and thus foster behaviour for learning.

In conclusion

Planning lessons can be time-consuming, and requires a huge amount of intellectual work as we draw on all our professional knowledge as teachers to consider how to help our students to learn. This chapter has drawn on a number of theoretical ideas in order to unpick what lesson planning involves, its purpose and why it matters, taking as the starting point 'who are my students and what do I want them to learn?'. Spending this time reading and developing our understanding of lesson planning will hopefully enable us to be more focused and strategic in our own planning, or as we support other teachers with their planning.

Further reading

Driver, R., Squires, A., Rushworth, P. & Wood-Robinson, V. (1994) *Making Sense of Secondary Science*. Abingdon: Routledge

Reiss, M. (2011) *Teaching Secondary Biology*. Hatfield & Banbury: Association for Science Education and Hodder Education

Sang, D. (2011) *Teaching Secondary Physics*. Hatfield & Banbury: Association for Science Education and Hodder Education

Taber, K. (2012) *Teaching Secondary Chemistry*. Hatfield & Banbury: Association for Science Education and Hodder Education

References

Cooper, P. & McIntyre, D. (1996) *Effective teaching and learning: Teachers' and students' perspectives.* Maidenhead: Open University Press

Harlen, W. (Ed.) (2015) *Working with Big Ideas of Science Education.* Trieste: IAP SEP. Available at: www.ase.org.uk/documents/working-with-the-big-ideas-in-science-education/

Kind, V. (2009) 'Pedagogical content knowledge in science education: Perspectives and potential for progress', *Studies in Science Education,* **45,** (2), 169–204

Krathwohl, D.R. (2002) 'A revision of Bloom's Taxonomy: an overview', *Theory into Practice,* **41,** (4), 212–218

Loughran, J.J., Berry, A. & Mulhall, P. (2006) *Understanding and developing science teachers' pedagogical content knowledge.* Rotterdam, Netherlands: Sense

Powell, S. & Tod, J. (2004) *A systematic review of how theories explain learning behaviour in school contexts.* London: University of London

Rowland, T., Turner, F., Huckstep, P. & Thwaites, A. (2009) *Developing primary mathematics teaching: Reflecting on practice with the Knowledge Quartet.* London: Sage

Shulman, L. (1986) 'Those who understand: knowledge growth in teaching', *Educational Researcher,* **15,** (2), 4–14

Shulman, L.S. (1987) 'Knowledge and teaching: Foundations of the new reform', *Harvard Educational Review,* **57,** (1), 1–22

Summers, M., Kruger, C. & Mant, J. (1998) 'Teaching electricity effectively in the primary school: a case study', *International Journal of Science Education,* (57), 1–22

CHAPTER 9

Inclusive science education

Lucy Dix, Fiona Woodhouse & Indira Banner

This chapter explains different ways of thinking about inclusion and offers a broad description of what inclusion may mean in school and how inclusion could be conceptualised within the classroom. It offers some practical ways of how science teachers might address a more competency approach to inclusion and suggests methods to plan and teach lessons with all students involved and benefiting from an inclusive approach.

Introduction

Inclusion means thinking about involving *all* students in the class and this chapter looks at teaching students who have special educational needs and disabilities (SEND) and those who have English as an addition language (EAL). Taking a competency approach to inclusion will open up a range of possibilities when working with *all* the students in the science classroom, with all their glorious diversity and difference.

Special Educational Needs and Disabilities: An overview

The term 'inclusion' is used in a variety of ways according to the particular conceptualisation of disability and 'special needs'. A rights-based understanding of inclusion is born from the rejection of the medicalisation and pathologising of disability. This 'medical model', also usefully termed the 'individual tragedy model', has its roots in the early 1800s' conceptualisation of impairments as something that could be fixed or, at the very least, remediated with the help of medical professionals. Along with the newly developed discipline of psychology, medical professionals saw impairments (bodily or cognitive) as, at best, a problem that could be fixed or, at worst, a reason to institutionalise large groups of the population. The legacy of this medicalisation is that we continue to conceptualise disability in these terms.

When considering the needs of students in our classrooms, we need to be clear regarding our inherent assumptions about the value of normalcy, the desirability of becoming less disabled and what we consider to be important factors in a good quality of life. It is important to consider whether the choices we make are based on our non-disabled (ableist) assumptions that, for example, taking a student out of class for more literacy instruction will necessarily improve his/her quality of life. There is a realistic argument that suggests that learning to read will give a child more life chances, but this needs to be balanced with the potential harm of removing the student from peer relationships, social learning and the knowledge that they 'belong' in the mainstream classroom. From a rights-based perspective, we can see that, whilst students have a right to learn to read, they also have a right to be part of their classroom community, to develop friendships and have fun, all of which will impact substantially on their quality of life. .

In current Government SEND documentation, 'inclusion' is not clearly defined, but a general statement, that the *'UK Government is committed to the inclusive education of disabled children and young people and the progressive removal of barriers to learning and participation in mainstream education'* (p.25, see: https://www.gov.uk/government/publications/send-code-of-practice-0-to-25), suggests that inclusion is a situational and structural concept and is to be tackled by removing structural barriers (for example, access to classrooms, or providing adequate safe spaces). Whereas the Centre for Studies on Inclusive Education (CSIE) considers inclusion to be *'[…] founded upon a moral position which values and respects every individual and which welcomes diversity as a rich learning resource'* (http://www.csie.org.uk/inclusion/index.shtml). Whilst the former 'barriers to learning' approach may be helpful in creating policies and practices to remove these barriers, a clear moral position such as that articulated by the CSIE is much more useful when considering inclusive practices in the classroom. If inclusion is reduced to removing a series of barriers, then the supposition may be that, once these barriers are broken down, inclusion is successful. The moral position taken by the CSIE suggests that inclusion is not just a case of modifying space, tasks or policies, but also the active promotion of inclusive thinking, which should be integral in the community of our classrooms, for teachers and students alike. Teachers and students who recognise and value diversity in their classrooms support the idea that 'inclusion' is not just about what happens in schools, but is about a wider social imperative.

To help move away from the individual and medical models of disability, we can examine a wider narrative, which considers how the role of societies and structures within societies restricts those with impairments, thereby rendering them disabled. Termed the 'social model' of disability, this conceptualisation grew from a rising acknowledgement from disabled people that it was not their impairments that were preventing them from being able to fully participate within society, but structural features (both physical and conceptual) that prevented access, created barriers and marginalised them. Although this conceptualisation is very useful in our schools as it encourages us to consider how our institutions may create physical barriers to disabled students' participation, it is necessary to be aware that conceptual barriers can be as disabling as the physical.

To illustrate how simply using a 'removing physical barriers to learning' approach can marginalise disabled people, let's consider a child who uses a wheelchair. Although she attends a mainstream school, she is required to use the back entrance to the school, because this is the only place where a ramp could be installed. Not only is she forced to use an entrance away from her peers, away from the playground games and chatter wherein those all-important friendship bonds are built, it is an entrance at the back near the dustbins. She is asked to arrive at school early so that a staff member can help her get ready for class, so she is always first in. She gets to class early, for the convenience of the school, so she always begins the school day on her own, waiting for the rest of the class to come in together through the main classroom door, highlighting her physical separation. It is not only the physical barriers that will affect this girl, there are attitudinal and conceptual barriers too; do the teachers or students question why provision is not made for her to come in the same door as her peers, or do they accept her separations as normal and inevitable? How is the girl to make friends or to participate in a social life if she doesn't have access to the same space in the morning? Is the lack of equality in her situation raised and questioned by the school on a regular basis?

Once in the science classroom, the girl cannot sit at the bench, unless the school has benches that change height to allow the wheelchair in place. This means that she will have difficulty doing any practical work and so her opportunities to engage in science activity are reduced to a spectator role.

A disability studies perspective

To be inclusive practitioners, we need to question these deeply rooted assumptions about 'what is best' and which lead to choices being made on behalf of students with disabilities. In this sense, we have a collective responsibility to challenge approaches to inclusion that appear to be moving away from a medical model, because the student is not in a segregated school or class. It is difficult for non-disabled people to recognise that not only are these disablist notions not being challenged, they are often perpetuated by everyday choices. One of the difficulties for staff in schools is that the foundations upon which disabled students are 'included' in mainstream schools appear progressive because a 'removal of barriers to learning' approach is taken. Whilst appearing to be based on a social model of disability, which considers how school structures create barriers for the student, the output of these approaches (Individual Education Plans – IEPs and Education Health and Care Plans – EHCPs, for example) relate back to what is problematic *for the student*. For example, a student who has moderate learning difficulties may have EHCP targets such as:

1. Move to level 4 in reading by the end of autumn term (out-of-class reading sessions 3 x a week).
2. Develop strategies for consistent spelling of high frequency science words (spelling practice with Teaching Assistant (TA)).
3. Recognise dual meaning words in science and be able to describe them (work with TA on creating a science dictionary).

Such targets highlight the students' 'inability' or 'difficulty' with certain tasks and make it their responsibility to practice and 'get better' (or more normal) at these skills.

To move away from a 'barriers to learning' approach, we can look at taking a competencies and rights-based approach. This moves away from considering a disabled student as unable to do something because of individual, structural or social barriers and considers each student as an individual with agency, someone who is in charge of and knowledgeable about his/her own life. In this approach, disabled students are seen as informed about their own lives and as competent decision-makers. Whilst social and structural barriers need to be continually questioned and dismantled, a competencies approach affords a disabled student participatory and decision-making rights. Central to this approach is the recognition that a student's abilities and competencies should be considered before choices are made about what and where s/he should be learning.

Taking a competencies approach to the targets set above would mean firstly asking the student what they feel may be preventing their learning, and what they think might help them to learn in a suitable way. The focus is less about what the student has to do to 'be better at' and more about what can happen to build on his/her current competencies. The student has a right to be able to access the material s/he needs in order to develop further competencies. The targets may look something more like this:

1. To find a wider range of reading materials in the library (or from home) and to share these with a teacher or friend.
2. Ask all teachers to display high frequency scientific words in their classrooms so that they are easily referenced.

3. Create a class- (or group-) generated book (or eBook), which explains dual meaning words in science and that is available in all science lessons.

English as an Additional Language: An overview

Students who are exposed to a language at home that is known or believed to be other than English are recorded as having English as an Additional Language (EAL). The recent census data (January 2017) indicated that there are 8.67 million students in schools in England, 4.7 million in primary and 3.2 million in secondary schools. Within these figures, 20.6% of primary students and 16.2% of secondary students are classified as students with EAL (see: https://www.gov.uk/government/uploads/system/uploads/attachment_data/fil e/650547/SFR28_2017_Main_Text.pdf).

This EAL classification is not a measure of students' proficiency in the English language or of their science knowledge or understanding; however, working with students with EAL requires careful planning for within our science lessons. It is important to establish, firstly, the student's level of English proficiency and, secondly, their abilities in science and what science they may have been previously exposed to, particularly if they are joining secondary provision from abroad. Students with EAL may:

➢ Have progressed through the UK school system and are at the correct stage of cognitive development, having participated in the current system but have a different language spoken at home;

➢ Be recent arrivals into the UK school system and have had little or no previous educational experience;

➢ Have had an age-appropriate educational experience in their mother tongue, but have limited exposure to the English language;

➢ Have a good level of exposure to the English language, but a limited formal education; and/or

➢ Have fluency in the English language, and are multilingual, and have good science knowledge.

As well as coming with different experiences and competencies, there will also be a cultural factor. It may be worth considering the way that science is taught in your school, because the use of practical investigations and also the curriculum content vary greatly between countries, giving students different experiences of science. It is important to recognise that speaking multiple languages and having a different culture is something to be celebrated;

students can teach us things from their first language and from their culture. Some teachers feel that students speaking in their first language should not be allowed or is not beneficial for students. In fact, it lowers the cognitive demand on tasks for the student to think in their own language until their English is stronger and, in schools where there are EAL students who speak the same mother tongue, it may be beneficial to place these in the same class. Hearing science ideas translated into English will help both their science and English learning.

Competencies-based approach to teaching and learning in science

Inclusion needs to start at the planning stage, so that a teacher is planning for all students to be included in the learning in classrooms. This means that we need to plan for what is going to be learned, how it is going to be learned and how students will show what they have learned.

Something as simple as making sure that all students can at least access the learning objectives and should be able to meet (some of) the learning outcomes is a good start. With these in place, we can then think about how all the students are going to achieve these outcomes. Aspects such as seating plans and organisation of the learning space should be arranged so that students' learning (and not behaviour, for example) is supported. If some students need to be near to the front to be able to read the board, for example, then others should be alongside them so that students have partners with whom to do practical work and to discuss their ideas. Using group work effectively is often a challenge for teachers and can be used to support all students' learning. Be careful, when assigning roles to the members of the groups, to ensure that all students have the opportunity to assume all types of roles across a year and that certain students are not excluded from roles that may be challenging.

Planning teaching to involve a range of activities should allow different students at different times to show their strengths. Use of media other than pens and whiteboards is often helpful, so that students can present their ideas in pictures, poems, mock interviews, role plays, etc. Having open-ended tasks can also support all students in doing their best: if asked to write a list of metallic elements, for example, it is unlikely that any student would be able to complete such a task. Using reference books and tablets to do research is an important skill: don't avoid this kind of work for students with weaker literacy or English language skills; instead, ask them what they need to make the work accessible (you may need to find them particular websites, give them keyword sheets, dictionaries or increase font sizes, for instance).

104

Teachers are expected to know which students in their class are on the SEND register and which have EAL and to understand what their support needs are, in order to ensure that these students are able to make progress in their classes. However, following procedures and policies within the school is not enough to effectively promote inclusion and challenge inequality. The approaches to inclusion that we will discuss here are part of a rights-based and competencies approach towards inclusion, one that moves inclusion from a 'barriers to learning' approach to one that recognises the self-determinacy, independence and competency of all students, regardless of label. One way we can begin is by asking disabled students and those with EAL about their own lives and the ways in which science benefits or harms in their country of origin, as well as in the UK. Listening to and including all students' opinions in classroom organisation and decision-making will help to ensure that our approach is based on rights and competencies, rather than outdated models of disability or ideas about language and culture. Do not be afraid of asking all students about their needs; an open and honest discussion about these will create the beginnings of an inclusive and respectful atmosphere in the classroom. Areas that students may want an input into could be:

- Seating arrangements
- What is displayed on the walls
- How PowerPoints and boards look (contrast/brightness/fonts)
- How groups are arranged
- How any teaching assistants or support staff are deployed
- How instructions are given

We could ask for student input by giving them some starter sentences to complete, such as:

- I wish you knew…
- Please can you…
- Other students could…
- I'd like to change…
- It would help me if…

All our students are competent people, but they are not identical; therefore, this approach will allow us to engage our students in a way that will tailor our teaching to their needs as a group, and as students with a variety of learning styles and needs.

Key strategies

Literacy

A main concern in science is often the literacy skills of the students and the requirements needed to access complex and specialist terminology. Students have literacy difficulties for a wide variety of reasons, including having EAL, and the strategies outlined below can be used to support many of them. Both the role of language in learning science and issues around literacy are discussed later in Chapters 16 and 17, but following are some issues to consider particularly in the context of inclusion.

> ➢ As we all know, there is a whole language that is specific to science and vocabulary development is essential to make progress. New and complex words in science can be difficult for all students, as can the dual meaning of words that students may have learnt in other or everyday contexts, but which also have a specific scientific meaning. Additionally, the grammatical structure of sentences that are used in science can be difficult, and it is important that we convey the relationships between words to give the precise meaning.
> Use of idioms in classroom talk might be understood literally and therefore be confusing to new speakers to English. The National Strategy unit of Literacy in Science (2002) has further support and offers three categories of scientific vocabulary (see below), which can be used to help students to understand and remember more of the extensive terminology:
> - ○ Names of objects (e.g. petal)
> - ○ Scientific processes (e.g. transpiration)
> - ○ Concepts (e.g. energy)

There are also synonyms that need clarification, such as 'skin' and 'membrane', and many words that have different meanings in science, such as 'cell' and 'tissue'. We need to firstly be aware of these issues in our classrooms and then plan to support learners. Some strategies to use would be using word walls of key terms for the topic, or vocabulary sheets with definitions and, if possible, translations (many smart phones and computers can now do this quickly).

Discussing the roots of words as well as meanings is important to help understanding and remembering. The students need to have opportunities to actively use the words to make sentences; doing this orally is important, as are directed activities related to text (DART); both of these will help students to gain confidence in using science words. There are many word games, loop games, matching pairs and describing words that can be used to consolidate learning and check for any misconceptions.

Displaying words and images in the classroom can support language learning, e.g. putting up complex and specialist terminology in simple and readable formats. This could be in the form of plain posters around the room, or laminated sheets on each table with key terms for that lesson or scheme. Where students are expected to follow a sequence, create (or get the students to create) a pictorial diagram showing the steps that they need to follow. This could also be made in a card sort format to encourage students to consider the order of steps. Images should always be accompanied by sentences or key words to develop students' vocabulary. Using videos to support students' literacy can offer an alternative to written instructions or concepts. Whilst students watch, give them a chart to complete, to tick when they have heard a concept or step talked about. Asking students to record their own videos explaining a concept can give the teacher insight into their progress; asking them to write up their script is a useful literacy activity that will encourage them to use scientific terminology.

To develop reading skills, active reading strategies can be used in which students are asked to highlight key ideas and key words on worksheets. Using writing frames or developing tables for students to complete to scaffold their learning further will also help. The teacher could take a photo of what they have done in terms of matching exercises or practical work, which can go into their books as a stimulus to begin writing. There are many ways to capture science other than in words, including diagrams, presentations, drama and recordings, all of which will support learning.

Academic English can take many more years in which to reach proficiency than social language – if this proficiency is not reached, this prevents students from attaining as highly as they can and as highly as native speakers (Hakuta *et al*, 2000). Separating English language support and science learning into 'two separate domains' means that they are often taught in separate classrooms and *'this undermines students' abilities to meaningfully access the content area texts'* (Hakuta *et al*, 2000 p.204; also see Lee & Buxton, 2013).

Practical skills in science and managing risk

What makes science different from many other subjects is the role of practical work. Many students with SEND and those with EAL may benefit from seeing concrete examples of abstract concepts. Both disabled students and recent immigrants may not have much experience doing practical work, so it may be useful to think about initial intervention work with them. Using health and safety sheets that are in the student's mother tongue or with pictures in addition to words will help to ensure that they are safe. Particularly if the students join mid-year in secondary school, the intervention could take the form of using teaching assistants or possibly 6th formers to do some simple health and safety work, lighting Bunsen burners, or assembling apparatus. This will give the students confidence with basic apparatus. Having pictures as well as words on cupboards and doors also help with identification of equipment. Do practical schedules have to be written? Could they be flow diagrams instead? Would this help? Think about group structures for practical work and designing roles within these to integrate, but also provide learning through examples. Good resources in which to engage come from CLEAPSS, to support safe working in practical lessons.

There are very few legitimate reasons as to why not all students in a class can take part in all practical activities. An important aspect of teaching experimental science is engaging the students in planning and preparing their own inquiries. Part of this preparation should include a discussion of how they think all students can be involved and what each student might need to allow them to fully participate. Make sure, however, that children with SEND are suitably challenged and not always given low-risk, low-challenge roles. Generating this kind of discussion in the classroom will create an open and supportive atmosphere. Teachers might include a range of activities where students choose, or are directed, to participate in a certain number of them; or students may each do a similar practical but with slight changes in equipment or measurements so that students are challenged appropriately; it could be that groups of students do their own practical and everyone reports back to the class about the one that they did. Giving students the choice as to what task they want to do is an inclusive measure and can be very effective for learning.

Often a main concern for teachers is that of managing risk. This might mean that we try to avoid certain practical activities with some students, perceiving the risk to be too high. However, there is risk for all students and, if we take a competency approach, students can be supported to manage this risk. Teachers should complete risk assessments with students to discuss and minimalise risks; it is rare that the risk can be removed completely for any student. An obvious 'risky' activity in our science classrooms is lighting a Bunsen burner, a task that we may consider to be too risky for a student with, for example, visual impairment or fine motor difficulties. To address this risk, first consider a rights-based approach: does the child have a right to take part in this learning? Then consider a competencies approach: what can the student do, what is she good at, what are her strengths? Key to answering these questions is consultation with students. We may find, on asking, that a visually impaired student has an acute sense of heat, and so will be able to light a Bunsen burner with minimal risk. A student who has poor fine motor skills may ask that they have a larger splint to hold (such as a candle), or that the splint be modified by having a plasticine grip made (which may benefit many students).

Recording practical work

Practical work can be recorded in a variety of ways to support the range of learners in the classroom. Students could record a discussion between the group, video their practical and provide a commentary, create a 'comic strip' of their practical work (on paper or using an App) or take photos and use them as prompts for written descriptions. As with many inclusive practices, this is likely to benefit all students, as they have to think about the purpose and learning objectives of the practical, which is an area often missing in classroom practical work (see Chapter 12).

Using resources and textbooks

We are fortunate in science in that much of our teaching is through the use of models and visuals; this is an important strategy to embed into all aspects of lessons to support all pupils as well as those with SEND or EAL needs. With access to the Internet in many classrooms, there are many resources that we can use to support teaching. Simple activities such as putting diagrams in the correct order will help the teacher to assess what the students have understood about processes in science.

Textbooks are now more illustrative and many have versions at different levels, which can be used to address concepts in a more accessible language. There may be books written in the student's mother tongue that s/he could use alongside the English equivalent to understand the science input.

Adult support

Often students with SEND and, to a lesser extent, students with EAL have a teaching assistant assigned to support them. While many teaching assistants are good, they are not usually trained science teachers and may not have any specialist knowledge in science or pedagogy; therefore, support for specialist knowledge should be provided by the teacher, allowing the teaching assistant (TA) to work with other students in the class.

TAs often come to a lesson without any prior knowledge of what will be taught; handing the TA a copy of the lesson plan will allow them to support students more effectively. When planning sessions, ask both the TA and the student how they would like the TA's time to be used. The student may feel competent in some aspects of the lesson, in which case the TA can be deployed elsewhere to support other students.

In conclusion

All the suggestions above follow the same two questions; does the student have a right to this knowledge and what are the student's current competencies (and how do we build on them)? At the heart of inclusive practice is recognition that all the children in our classrooms have a right to learn with their peers, to develop social as well as academic competencies and are inherently experts about their own needs.

Inclusion in the science classroom is much more than an understanding of policies or knowing a child's EHCP; it is a recognition of the damaging conceptualisations of disability and students who do not speak and understand English fluently, which are pervasive within school systems; it is a constant questioning of our own assumptions; it means affording all children the same rights; and it means actively engaging with a competencies-based approach.

Some resources to support SEND

The books below offer general insights into SEND:

Bates, B. (2017) *A Quick Guide to Special Needs and Disabilities*. London: Sage

Hudson, D. (2016) *Specific Learning Difficulties*. London: Jessica Kingsley Publishers

The websites below are also a good starting point:

http://www.csie.org.uk
The Centre for Studies on Inclusive Education provides guidance and support.

See https://www.gov.uk/government/publications/send-code-of-practice-0-to-25 for the current SEND Code of Practice (0 to 25 years)

ASE has some ideas to support the teaching of pupils with SEND: https://www.ase.org.uk/resources/send/

http://www.cleapss.org.uk/
This website has a great deal of practical support for the teacher in the classroom.

Some resources to support EAL

Ardasheva, Y., Norton-Meier, L. & Hand, B. (2015) 'Negotiation, embeddedness and non-threatening learning environments as themes of science and language convergence for English language learners', *Studies in Science Education*, **51,** (2), 201–249. DOI: 10.1080/03057267.2015.1078019

Braden, S., Wassell, B.A., Scantlebury, K. & Grover, A. (2016) 'Supporting language learners in science classrooms: insights from middle-school English language learner students', *Language and Education*, **30,** (5), 438–458, DOI: 10.1080/09500782.2015.1134566

DfES (2005) *Aiming High: Guidance on the assessment of students learning English as an additional language,* HMSO. Retrieved from: https://www.naldic.org.uk/Resources/NALDIC/Teaching%20and%20Learning/5865-DfES-AimingHigh1469.pdf

DfES (2002) *Key Stage 3 National Strategy Access and Engagement in Science.* Retrieved from: https://www.naldic.org.uk/Resources/NALDIC/Teaching%20and%20Learning/0610-2002Science.pdf

Fernando, P. & Cooper, R. (2017) 'Teaching Strategies: Supporting EAL students in learning biology terminology', *Teaching Science,* **63,** (1), 34–40

The websites below should be helpful:
https://naldic.org.uk/
The National Association for Language Development In the Curriculum (NALDIC)'s website has current data as well as a host of advice and support.

http://www.elsp.ie/indexLS.shtml
The English Language Support Programme developed by Trinity College Dublin (ELSP) has some useful science support resources – see:
http://www.elsp.ie/science.shtml
https://www.theealacademy.co.uk/team/kamil-trzebiatowski/

There are several translation programmes that will enable the translation of words and phrases and some will also do text-to-speech. This will help create resources and additionally make the student feel welcome within the classroom. Bilingual dictionaries, where available, would also be invaluable. The aim, however, is to enable the student to become confident with the English language.

References

Centre for Studies on Inclusive Education (n.d.). Available at: http://www.csie.org.uk

DfE (2014) *SEND Code of Practice 0 to 25 years.* Available at: https://www.gov.uk/government/publications/send-code-of-practice-0-to-25

DfES (2004) *Key Stage 3 National Strategy: Literacy in Science.* HMSO/DfES 0561-2002

Hakuta, K., Butler, Y.G. & Witt, D. (2000) *How Long Does It Take English Learners To Attain Proficiency?* USA: Stanford University. Available at: https://eric.ed.gov/?id=ED443275

Lee, O. & Buxton, C.A. (2013) 'Teacher professional development to improve science and literacy achievement of English language learners', *Theory Into Practice*, **52,** (2), 110–117. DOI: 10.1080/00405841.2013.770328

United Nations (n.d.) *Convention on the Rights of Persons with Disabilities (CRPD)*. Available at: https://www.un.org/development/desa/disabilities/convention-on-the-rights-of-persons-with-disabilities.html

CHAPTER 10

Equality and diversity in science education

Clare Thomson

The diversity of our science classroom is to be celebrated and should be used as an asset to help students' learning. However, persistent under-achievement and under-participation by a particular gender, social class or ethnic group(s) is of concern to many teachers. This chapter presents an overview of current thinking as to why this under-achievement and under-participation occurs, and what actions teachers and schools can take to improve equality and diversity in school science.

Introduction

'Despite any political or popular consensus over what science education is for (e.g. creating the next generation of scientists versus producing a scientifically literate population (Millar, 2014)), there is broad agreement in many quarters around the importance of increasing participation in science once it is no longer compulsory, particularly among groups who have been historically under-represented in science' (DeWitt & Archer, 2015 p.2171).

In Britain, the predominant option for post-compulsory participation in science is via A-levels or Scottish Highers. Those students pursuing science at A-level tend to be high attaining, mainly scoring the top two grades in national examinations such as GCSE, and from higher social class backgrounds. While there are strong links between socio-economic status (SES) and attainment, participation in science post-16 also correlates with ethnicity and gender in an intersectional way.

However, the Equality Act 2010 puts a responsibility on schools to ensure that no students are discriminated against with particular regard to the protected characteristics of: disability, gender reassignment, race, religion or belief, sex and sexual orientation. Advancing equality of opportunity involves in particular:

> ➤ removing or minimising disadvantages suffered by people that are connected to a particular characteristic that they have (for example, disabled students, or gay students who are being subjected to bullying);

> taking steps to meet the particular needs of people who have a particular characteristic (for example, organising a fieldwork day to enable Muslim students to pray at prescribed times); and

> encouraging people who have a particular characteristic to participate fully in any activities (for example, encouraging both boys and girls, and students from different ethnic backgrounds, to be involved in science clubs).

This chapter focuses on the three areas of diversity: socio-economic status (SES), gender and ethnicity, and explores what can be done to make school science as inclusive as possible and to ensure genuinely equal access for all students. For each of these, data will be presented to demonstrate the extent to which students' attainment and participation in science is affected by these factors, followed by a discussion of recommendations for teachers and schools from a variety of sources.

Attainment data on England's school students are more extensive in coverage, detail, quantity and accessibility than those of many other European countries. These data facilitate investigation of attainment in England and its relationship to ethnicity, gender and poverty. Analysis of longitudinal sample studies shows recurrent correlations of low attainment with specific ethnic minority groups, with gender and, most strongly, with sections of society on a low income.

The Sutton Trust report (Jerrim, 2017), which used 2015 OECD Programme for International Student Assessment (PISA) data for 16 year-olds, found that the top 10% of students in England had a PISA test score of 642, but the top 10% of low-SES students in England had a PISA test score of 598 (see Chapter 20 for more discussion of PISA and other international assessments). This gap corresponds to the highest attaining low-SES students being about 2 years and 8 months' school years behind the top 10% of students overall – approximately the OECD average. This gap also has a gender aspect: the average test score for low-SES boys is 71 points below the average test score for boys, but the average test score of low-SES girls is 93 points below the average test score for girls.

Socio-economic status

Free School Meal (FSM) rates vary enormously. These represent the proportion of students who are eligible *and* are claiming FSMs. While most ethnic minority groups have higher *rates* of FSM, a greater *number* of FSM children are White (over 700,000). It is sometimes argued that the rate of FSMs is of limited use in understanding fully the relationship between gradations of poverty and attainment, and so Income Deprivation Affecting Children Index (IDACI)[1]deciles, which are calculated for small areas within a local authority, can be more useful.

The Joseph Rowntree Foundation report, *Falling Short: the experiences of families living below the Minimum Income Standard* (p. i) concludes that:

➢ *'families need stability, but this is undermined by irregular employment and hours, changes in benefits and tax credits, and insecurity in private rented housing;*

➢ *'coping on a low income involves constant monitoring of budgets, hard work and discipline, but the stress of trying to keep on top of finances is emotionally draining;*

➢ *'parents tend to prioritise meeting their children's needs and sacrifice their own'.*

The impact on children includes their access to paid-for after-school activities and to holidays and day trips, which broaden children's perspectives on the physical and natural world. It cannot be assumed that parents and children from low SES backgrounds have low aspirations when thinking about progression to science-related careers. Recent research (Alcott, 2017) using the DfE Longitudinal Study of Young People in England (LSYPE) data set suggests that a teacher's interest and support has a positive impact on a student's likelihood of progression to university. Students whose academic achievement falls in the middle third and whose parents lack higher qualifications themselves benefit most from this encouragement.

Ethnicity

Nearly 80% of students in schools in England are White, and over 90% in that group are White British. The largest single minority ethnic group is Pakistani, followed by Black Africans, the latter having more than doubled in number over the last 10 years. No individual minority group constitutes more than 4% of the total school population, but they are not evenly spread across the UK. Some local authorities have over 50% ethnic minority students in their schools (Parsons, 2016).

A recent Sutton Trust research brief (Kirby & Cullinane, 2016) reports that, historically, ethnic minority students have faced the greatest educational challenges in the UK, but that today these patterns have been at least partially improved, as many ethnic minority students have seen improved results relative to the national average. Several explanations have been proposed for this shift – the popularity of private tutors providing extra support for students amongst ethnic groups, and differing levels of parental aspiration, amongst others.

[1] The Income Deprivation Affecting Children Index is the proportion of all children aged 0 to 15 living in income-deprived families.)

However, while ethnic minority students eligible for FSM have seen gains relative to their White British counterparts at school, they often face challenges in post-secondary education. Ethnic minorities now enrol in university at rates higher than their proportion of the national population, but fewer attend top universities and enter full-time employment after graduation. Every ethnic minority group, apart from Gypsy/Roma, attend university at a higher rate than White British. But, of those who enrol at university, most minority groups rank below White British when it comes to attending more selective institutions (UCAS, 2016). Indeed, the most advantaged students are six times more likely than the most disadvantaged students to study at the more selective universities (Wyness, 2017) (See Figures 1 & 2).

Relatively few studies have explored the ways in which minority ethnic young people experience and participate in science, but those that have demonstrate that there are diverse experiences of, and approaches to, science among them (Archer et al, 2015; Wong, 2016). To explain the differences in participation, Archer et al (2015) proposed the notion of 'science capital' – a conceptual device drawing together all the science-related resources and dispositions that an individual holds. Wong's qualitative study empirically mapped out key characteristics of science participation (2016), suggesting that 'science adverse' students have below-average levels of achievement, low interest in and low science capital, expressing no aspirations toward science-related careers. In his study, these were mostly Black Caribbean boys and girls (n=9). 'Science intrinsic' students express science-related career aspirations and some levels of science interest and science capital, despite below-average attainment in science (mostly Bangladeshi boys in his study, n=8). 'Science intermediate' students came from all five minority ethnic backgrounds in the study (n=6) and expressed at least one science career aspiration, having average achievement, science interest and science capital. 'Science extrinsic' students were mostly Chinese boys and girls (n=13) in this study, with no expression of science interest or science career aspirations, but having average or above average achievement and science capital. The last group, (mostly Indian boys and girls, n=10) were 'science prominent' students with above-average science achievement, high levels of interest and high science capital, and expressed science-related career aspirations.

However, their science career aspirations are likely to be medical-related, i.e. a career *from* science rather than a career *in* science. So, while minority ethnic students are all under-represented in the study of physics at doctorate level in UK universities, they are generally over-represented in the study of medicine and dentistry (Woolf, Potts & McManus, 2011).

Figure 1: Attainment gap between % of non-FSM and FSM girls achieving 5A*CEM

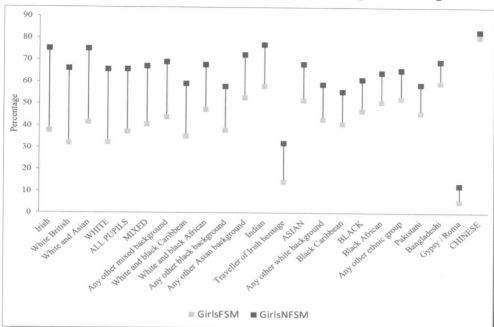

Figure 2: Attainment gap between % of non-FSM and FSM boys achieving 5A*CEM

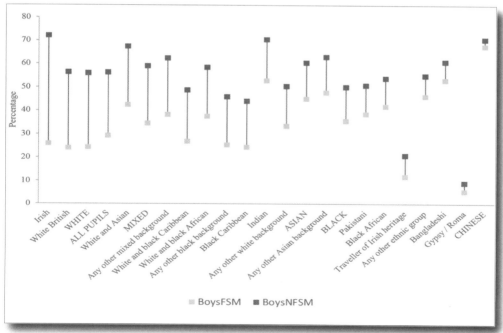

Source: The Sutton Trust Research Brief, November 2016, *Class differences: ethnicity and disadvantage*

Gender

Schools have rightly focused on ensuring that students are not disadvantaged by their sexual orientation and that name-calling and homophobic bullying are visibly dealt with. However, initiatives like the *Everyday Sexism* project clearly show that we are a long way from living in a gender-equitable society.

Within science and science-related subjects, career choices tend to follow a gendered pattern, with females choosing medical, health and biology careers and males favouring engineering and computer sciences. The physical sciences – in particular physics and engineering, continue to attract only a very small proportion of women post-16. This is despite many interventions over the last 15-20 years. Having rejected a suggestion of innate genetic differences, Blickenstaff (2005) suggested 8 possible reasons for girls' lack of engagement with science:

1. Girls' lack of academic preparation for a science major/career.
2. Girls' poor attitude toward science and lack of positive experiences with science in childhood.
3. The absence of female scientists/engineers as role models.
4. Science curricula are irrelevant to many girls.
5. The pedagogy of science classes favours male students.
6. A 'chilly climate' exists for girls/women in science classes.
7. Cultural pressure on girls/women to conform to traditional gender roles.
8. An inherent masculine world view in scientific epistemology (p.371–372).

As discussed by Blickenstaff, and by Osborne, Simon and Tytler (2009), this is clearly a complex problem that is not easily solved. The ASPIRES 2 project found that, for the seventy 15- and 16 year-olds and sixty-two parents who they interviewed in 2015, physics continues to appear as a subject for men. The lack of representation of women in physics (both in reality and as presented in popular media – think *The Big Bang Theory*) encourages the assumption that most women are unable to, or unsuitable for, work in physics. In addition, the content of the physics curriculum doesn't match what girls are interested in learning about science (Jenkins & Nelson, 2005), but girls would be more interested if they could see how physics relates to people (Krogh & Thomsen, 2005). The mathematical aspect of physics can also be offputting, as girls tend to be less confident in mathematics than boys (Lubienski *et al*, 2013). This, in conjunction with the under-representation of women, means that many young women are left feeling daunted by physics because they perceive it to be such a masculine subject and therefore not for them (Francis *et al*, 2017 p.170). Despite the emerging view of both parents and students that 'anybody can do anything', the notion that 'girlyness' and science cannot co-exist remains (Archer *et al*, 2013 p.3).

One potential reason underpinning such discrepancies, beyond associations between some science careers and masculinity, concerns the kinds of values that students seek to fulfil in their subject and career choices, whether they are related to knowledge and skills or more focused on working with people. Students with strong beliefs in their abilities in science are more likely to pursue science once it is no longer compulsory and to aspire to science-related careers. Girls' beliefs in their capabilities in science are often less positive than those of boys (DeWitt *et al*, 2013). They may be affected by the extent to which they receive recognition of their science achievements or encouragement to pursue science from teachers, family and friends, with research finding that females often receive less recognition and encouragement than males (Mujtaba & Reiss, 2013).

Using National Pupil Database (NPD) data, the Institute of Physics (IOP) found striking differences in the uptake of physics post-16 for girls and boys, depending on the type of school that they attended (IOP, 2012). Girls were almost two-and-a-half times more likely to go on to do A-level physics if they came from a girls' school rather than a co-ed school (for all types of maintained schools in England). The positive effect of single-sex education was not replicated in the other sciences. This suggests that gender-specific cultural influences were coming into play and that, in many schools, expectations of students were, and often still are, gender-stereotyped. Further research showed the extent to which gender stereotyping occurs in schools and how it creates barriers to subject choice post-16, for both boys and girls (IOP, 2013).

The whole school perspective

It can be argued that the causes of low attainment lie largely outside school and could be better tackled by interventions to lift people out of poverty, rather than adjusting in-school factors (Parsons, 2016). However, from the perspective of an individual school and teachers within it, knowing what might be effective ways of minimising the impact of poverty is important. Research using university entry data from the Higher Education Statistics Agency (HESA) showed that only 19% of students from the poorest fifth of areas enter Higher Education (HE), compared with 45% of those from the richest fifth of neighbourhoods (Sammons *et al*, 2016). It also found that, for students to fulfil their potential post-16, both attainment and aspirations matter, particularly for poorer students, so it is vital that schools and teachers see the value of promoting both self-belief and attainment as mutually reinforcing outcomes.

Effective in-school measures to raise attainment in science

Science courses at A-level and beyond require a high level of previous attainment. For example, prior attainment in science and mathematics is the strongest indicator of whether or not a student will go on to study physics. As prior attainment is strongly linked to SES, students from lower SES backgrounds are less likely to achieve the grades necessary to study science subjects post-16. A recent review for the Educational Endowment Foundation (EEF) and the Royal Society (Nunes *et al*, 2017) shows that scientific reasoning and literacy skills are important mediators of the relation between SES and science learning. In addition, interventions that supported the development of students' scientific reasoning and/or their literacy skills in science showed an improvement in science attainment for low SES students, in some cases an increased improvement compared with other students (Hanley, Slavin & Elliott, 2015). When looking at ways of improving attainment, the EEF teaching and learning toolkit is an accessible summary of educational research and gives an indication of the impact in terms of progress for each initiative. They suggest that feedback, metacognition and self-regulation can have the greatest impact on students' progress, in addition to the two factors mentioned above. Evidence shows that a number of organisational measures may also be effective in improving attainment for all students (IOP, 2014; Ofsted, 2015). Schools should:

Implement evidence-based strategies that improve attainment, such as:

- ➤ Identify designated staff and governors to champion the needs of disadvantaged most able students;

- ➤ Ensure that teachers and leaders in Key Stage 3 (age 11-14) use information held by primary schools about students' learning and achievements in Key Stage 2 (age 7-11) effectively, so that the work provides the right level of challenge;

- ➤ Give Key Stage 3 equal priority with other key stages when allocating teaching staff to classes;

- ➤ Ensure that the sciences are taught by teachers of their own specialism where possible;

- ➤ Provide training to improve teaching, which can help to break the cycle of poor attainment in schools in deprived areas;

- ➤ Instigate peer-learning programmes for students to help increase attainment, build self-confidence and give students opportunities to develop presentation skills;

- ➤ Avoid setting; if that is not possible, take into account that prior attainment may be heavily dependent on social background; and

➢ Select students for science enrichment programmes based on potential and interest rather than solely on past attainment.

Develop home-school partnerships

Involving parents with school and their child's education can help to increase a child's attainment and aspirations. Developing a successful programme of parental engagement takes time and trust between all parties concerned. These suggestions are likely to benefit all students, but may be particularly helpful to students from disadvantaged backgrounds. Schools should:

➢ Consult the whole school community to create and implement a meaningful 'Home-School Agreement', which can be recognised on all sides;

➢ Integrate parental engagement into a whole-school approach, rather than as a 'bolt-on' activity, using flexible models of working in partnership to maintain a genuine two-way exchange;

➢ Tap into parents' needs and interests by creating comfortable environments and involve other members of the community;

➢ Provide well-structured programmes (e.g. homework clubs) with high-level support to reduce drop-out rates;

➢ Train all school staff about the best ways to engage parents with their child's education and give teachers adequate time to undertake this work; and

➢ Provide advice on how parents can help their children with homework and generally improve educational achievement.

Providing appropriate advice on routes through education

The *UCAS Undergraduate 2016 End of Cycle Report* uses multiple equality measures (MEM)[2] to identify those groups with the highest and lowest entry rates to higher education overall. Young people in the top MEM group 5 are nearly four times more likely to go to university than those in group 1. However, entry rates to higher tariff providers ('top' universities) are ten times greater for those in group 5 (24.5%) compared to those students in group 1 (2.3%) (UCAS 2016). Whilst the most important thing is that students study the course that fits their interests and aspirations best, in the location that suits them best, there is evidence that disadvantaged students don't always know the full range of choices available to them (Wyness, 2017).

[2] The multiple equality dimension measure considers area and income background, school sector, gender, and ethnic group combined.

Many students and families from disadvantaged backgrounds are not aware of the range of careers that exist in STEM areas and, equally important, in areas where STEM skills offer a significant advantage. Consequently, subject choices are made without proper information as to where they might lead and the economic consequences of particular choices are not made visible. Parents who were not educated in the UK, and/or do not have an HE qualification themselves, may need basic information on how the system works and the options available beyond level 2 qualifications, in order to advise and support their children appropriately. Schools should:

➢ Integrate awareness of skills development into mainstream teaching; students should realise that science skills are applicable across a wide range of careers;

➢ Ensure that careers advice and guidance starts early enough (before Year 9, age 14, in England) to be effective; is bespoke to the student and their current aspirations; concentrates on the next stage of choice; and includes parents;

➢ Implement a proactive approach in matching work placements with students;

➢ Provide information to students and parents on how to access higher education and which subjects are desired by universities, as well as information about bursaries and other financial help; and

➢ Provide students and parents with information about the appropriate subjects and courses to study at school that are required to access HE science and science-related courses.

Increasing science capital

Science capital is a concept that can help us to understand why some young people participate in post-16 science and others do not. In particular, it helps to shed light on why particular social groups remain under-represented and why many young people do not see science careers as being 'for me'. As identified by Archer et al (2015), science capital refers to having science-related qualifications, understanding, knowledge (about science and 'how it works'), interest and social contacts (e.g. knowing someone who works in a science-related job). Students from under-represented groups will have few role models, people who are 'like them', visible in science or science-related careers. For such children, inspiration and information from adults who are not their parents, including teachers, can be important and influential. Girls especially are more likely to take a subject post-16 if they receive recognition of their science achievements or encouragement to pursue science from teachers, family and friends, with research finding that females often receive less recognition and encouragement than males (Mujtaba & Reiss, 2013). Schools should:

➤ Raise the overall profile of science in school. This requires support from senior leadership and the science department as a whole. Any strategies need to be embedded and should foster a general culture among adults in the school and the surrounding community of being positive about science STEM;

➤ Endeavour to build long-term relationships between students and role models (who could be ex-students) with a similar background in terms of geography and economic background. One-off visits are much less likely to be effective than establishing a successful science club;

➤ Make sure that all teachers are aware of the influence they can have on children's future careers; that they are informed about current entry routes to different careers and that they do not discourage students from pursuing science careers based on their personal opinions and stereotypes; and

➤ Explore socio-scientific issues in lessons: this has a positive effect on encouraging young people (especially females) to choose post-compulsory science education.

Figure 3: Working for inclusive teaching & learning in partnership

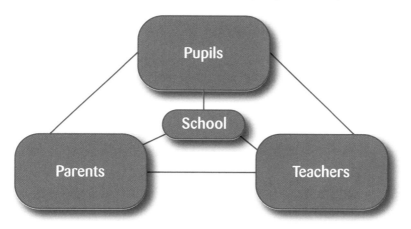

Improving gender balance

In addition to the advice set out above, to make a significant difference to students' perceptions of gender and subject choice, work needs to be done across the whole school to challenge stereotypes among both staff and students. Schools should:

➢ Identify one of the Senior Leadership Team (SLT) as a gender champion whose role includes bringing together the whole school in a coherent campaign to challenge gender stereotypes. Governors should be involved in the campaign in order to reinforce the message that this activity is a priority;

➢ Offer training to staff in equality and diversity awareness and unconscious bias, whether as part of their induction to the school or their ongoing professional development;

➢ Ensure that sexist language is treated as being just as unacceptable as racist and homophobic language and that gender-specific bullying and sexual misconduct are recognised on the school's reporting system;

➢ Collect and use gender-disaggregated data on both achievement and progression for all subjects, to be discussed formally at whole-school level, using benchmark data for comparison. Where there are issues to be addressed, actions are generated and targets are set;

➢ Introduce and develop initiatives on the basis of what works and in a way that shows how they address a problem identified in the school data;

➢ Use carefully planned visits and invite visitors into school to encourage students to challenge stereotypical views and engage positive role models who will commit to developing sustainable relationships with the school;

➢ Ensure subject equity with a strict policy that all subjects are presented equally to students in terms of their relative difficulty and encourage teachers to refrain from making any remarks about how difficult they found particular subjects;

➢ Encourage student ownership of the issue by making students the heart of any campaign to counter gender stereotyping; and

➢ Value personal, social, health and economic education (PSHE). Ensure that teachers are provided with good resources and activities and that sessions on equality and diversity form the basis of a wider school campaign, with discussions on these themes continuing through other topics.

The classroom perspective

In order to make science lessons, and physics lessons in particular, as inclusive as possible, a number of actions can be taken. It is worth bearing in mind that aiming for inclusive teaching that engages girls effectively is also good for boys. Using an inclusive learning checklist, such as the one provided by the IOP, can be a helpful first step. From a variety of research and intervention projects, it is suggested that teachers should:

➢ Ask a supportive colleague or student to observe a lesson and monitor interactions and questions to or from boys and girls. This can be a powerful eye-opener, even if teachers think they are mindful of gender balance. Evidence shows that boys tend to dominate lessons in terms of shouting out answers to questions and putting their hands up a great deal more (Archer *et al*, 2017);

➢ Manage practicals by assigning job roles that rotate regularly to ensure that everyone has a chance to both set up and use equipment and write down results;

➢ Actively manage groups, thinking about the purpose of each activity and whether single sex or mixed sex groupings will be most effective. Consider whether the groups should be mixed ability or set in some way. Don't allow the groupings to become fixed for all lessons, or use sitting with girls to manage the behaviour of difficult boys;

➢ Use culturally- and gender-aware contexts and examples – many examples commonly used in physics, in particular, have a male bias: football and F1 cars to name but two. Aiming for more gender-neutral examples or using a wider range of contexts helps all students to feel engaged with the subject. Thinking carefully about the backgrounds and experiences of the students in individual classes will help in the choice of appropriate contexts and examples;

➢ Make connections between topics explicit – for many students, and particularly for girls and physics, it is helpful to have the connections between what they have learned in the past and a new topic made clear. In addition, make explicit the transferable skills that they are acquiring in the process of learning science and how these might be useful in the future;

➢ Use inclusive teaching techniques – there are a number of different techniques, such as think-pair-share and 'plickers' (a question-response system using unique codes for students that a teacher can scan with a mobile phone (see https://www.plickers.com/)), which can be used to increase thinking time in class. These techniques also allow girls and quieter members of a class to participate more fully;

➤ Encourage students to be in charge of their own learning – use metacognition and self-regulation to help students learn to learn (see EEF toolkit for more information). Encourage students to adopt a 'growth mindset'. This is the name given by psychologist Carol Dweck (2017) to the idea that intelligence is not fixed and that effort leads to improvement and success;

➤ Ensure that gender equality and ethnic and cultural diversity are evident in displays in classrooms and corridors, whether of student work or aspirational role models; and

➤ Rethink science clubs – these can often be boy-heavy, which can put off interested girls and reinforce the idea that science is not for them. Research projects such as Cern@school and Crest Awards attract a better gender balance, as do science ambassador schemes, in which students do outreach in feeder primary schools. The IOP has resources to help with training students as science ambassadors.

A recent report by the EEF and Royal Society (Nunes *et al*, 2017) also recommended that teachers focus on developing students' reasoning and literacy skills.

Figure 4: Requirement for successful progression to STEM post-16

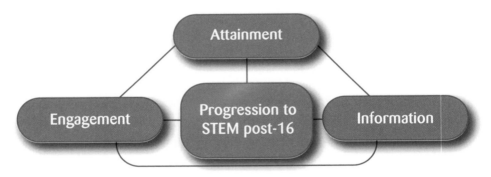

In conclusion

Many of the issues touched on in this chapter need addressing at the whole school level for significant long-term impact to be made. It is for that reason that a member of the school SLT as a champion of inclusion and diversity is so important. However, there are also relatively simple things that can be done by teachers in their classrooms, which can make a measurable difference to the way in which their students experience science and its role in their lives.

Further reading and resources

Archer, L. & DeWitt, J. (2017) *Understanding Young People's Science Aspirations – How students form ideas about 'becoming a scientist'.* Abingdon, Oxon: Routledge

Equality Act 2010 – for further information on how this applies in schools: http://www.legislation.gov.uk/ukpga/2010/15/notes/division/3/6/1

EEF Teaching and Learning toolkit: https://educationendowment foundation.org.uk/resources/teaching-learning-toolkit/

Gender balance – resources for both whole school action and the science classroom: www.iop.org/genderbalance

Institute of Physics (2015) *Opening Doors: a guide to good practice in countering gender stereotyping in schools.* London: Institute of Physics

Implicit bias – a test is available online via the Harvard University website at: https://implicit.harvard.edu/implicit/

Mindsets – for more information go to: https://mindsetonline.com/whatisit/about/index.html

Ofsted (2015) Examples of good practice in schools: https://www.gov.uk/government/collections/ofsted-examples-of-good-practice-in-schools

Science capital – for further exploration of this concept: http://www.kcl.ac.uk/sspp/departments/education/research/Research-Centres/cppr/Research/currentpro/Enterprising-Science/01Science-Capital.aspx

Sayed, M. (2011) *Bounce: The Myth of Talent and the Power of Practice.* London: Fourth Estate

Understanding Unconscious Bias – a short video from the Royal Society: https://www.youtube.com/watch?v=dVp9Z5k0dEE

References

Alcott, B. (2017) 'Does teacher encouragement influence students' educational progress? A propensity-score matching analysis', *Research in Higher Education*, DOI 10.1007/s11162-017-9446-2. Retrieved from: https://link.springer.com/article/10.1007/s11162-017-9446-2/fulltext.html

Archer, L., Dawson, E., DeWitt, J., Godec, S., King, H., Mau, A., Nomikou, E. & Seakins, A. (2017) 'Killing curiosity? An analysis of celebrated identity performances among teachers and students in nine London secondary science classrooms', *Science Education*, (101), 741–764

Archer, L., Dawson, E., DeWitt, J., Seakins, A. & Wong, B. (2015) '"Science Capital": A conceptual, methodological, and empirical argument for extending Bourdieusian notions of capital beyond the Arts', *Journal of Research in Science Teaching*, **52,** (7), 922–948

Archer, L., Osborne, J., DeWitt, J., Dillon, J., Wong, B. & Willis, B. (2013) *ASPIRES – Young people's science and career aspirations, age 10-14.* Final Report. Available at: https://www.kcl.ac.uk/sspp/departments /education/research/aspires/ASPIRES-final-report-December-2013.pdf

Blickenstaff, J.C. (2005) 'Women and science careers: leaky pipeline or gender filter?', *Gender and Education*, **17,** (4), 369–386

DeWitt, J. & Archer, L. (2015) 'Who Aspires to a Science Career? A comparison of survey responses from primary and secondary school students', *International Journal of Science Education*, **37,** (13), 2170–2192

DeWitt, J., Osborne, J., Archer, L., Dillon. D., Willis, B. & Wong, B. (2013) 'Young Children's Aspirations in Science: The unequivocal, the uncertain and the unthinkable', *International Journal of Science Education*, **35,** (6), 1037–1063

Dweck, C.S. (2017) *Mindset – Updated Edition: Changing The Way You Think To Fulfil Your Potential.* New York: Ballantine Books

Francis, B., Archer, L., Moote, J., DeWitt, J., MacLeod, E. & Yeomans, L. (2017) 'The construction of physics as a quintessentially masculine subject: Young people's perceptions of gender issues in access to physics', *Sex Roles*, **76,** (3-4), 156–174

Hanley, P., Slavin, R. & Elliott, L. (2015) *Thinking, Doing, Talking Science: Evaluation report and Executive summary.* London: Education Endowment Foundation

Hill. K., Davis, A., Hirsch, D. & Marshall, L. (2016) *Falling short: the experiences of families living below the Minimum Income Standard.* York: Joseph Rowntree Foundation

Institute of Physics (2012) *It's Different for Girls: the influence of schools.* London: Institute of Physics

Institute of Physics (2013) *Closing Doors: exploring gender and subject choice in schools.* London: Institute of Physics

Institute of Physics (2014) *Raising Aspirations in Physics: a research review.* London: Institute of Physics

Jerrim, J. (2017) *Global Gaps: comparing socio-economic gaps in the performance of highly able UK pupils internationally.* London: The Sutton Trust

Jenkins, E. & Nelson, N.W. (2005) 'Important but not for me: students' attitudes toward secondary school science in England', *Research in Science & Technological Education,* **23,** (1), 41–57

Kirby, P. & Cullinane, C. (2016) *Class differences: Ethnicity and disadvantage.* London: The Sutton Trust

Krogh, L.B. & Thomsen, P.V. (2005) 'Studying students' attitudes towards science from a cultural perspective but with a quantitative methodology: border crossing into the physics classroom', *International Journal of Science Education,* **27,** (3), 281–302

Lubienski, S.T., Robinson, J.P., Crane, C.C. & Ganley, C.M. (2013) 'Girls' and boys' mathematics achievement, affect, and experience: Findings from ECLS-K', *Journal for Research in Mathematics Education,* (44), 634–645. DOI: 10.5951/jresematheduc.44.4.0634

Millar, R. (2014) 'Designing a science curriculum fit for purpose', *School Science Review,* **95,** (352), 15–20

Mujtaba, T. & Reiss, M.J. (2013) 'What sort of girl wants to study physics after the age of 16? Findings from a large-scale UK survey', *International Journal of Science Education,* **35,** (7), 2979–2998

Nunes, T., Bryant, P., Strand, S., Hillier, J., Barros, R. & Miller-Friedmann, J. (2017) *Review of SES and Science Learning in Formal Educational Settings,* London: Education Endowment Foundation and the Royal Society. Retrieved from https://educationendowmentfoundation.org.uk/ public/files/Review_of_SES_and_Science_Learning_in_Formal_Educational_ Settings.pdf Accessed 11.10.17

Ofsted (2015) *The most able students: an update on progress since June 2013*: No. 150034. Manchester: Ofsted

Osborne, J., Simon, S. & Tytler, R. (2009) *Attitudes Towards Science: an Update.* Presented at the American Education Research Association Annual Meeting. Retrieved from: science.teachingjournalclub.org/wp-content/ uploads/2012/06/Attitudes-Toward-Science-OsborneSimon-Tytler.pdf

Parsons, C. (2016) 'Ethnicity, gender, deprivation and low educational attainment in England: Political arithmetic, ideological stances and the deficient society', *Education, Citizenship and Social Justice,* **11,** (2), 160–183

Sammons, P., Toth, K. & Sylva, K. (2016) *Believing in Better.* London: The Sutton Trust

UCAS (2016) *End of Cycle Report 2016 – UCAS Analysis and Research.* Available at: www.ucas.com/corporate/news-and-key-documents/ key-documents

Wong, B. (2016) 'Minority Ethnic Students and Science Participation: A qualitative mapping of achievement, aspiration, interest and capital', *Research in Science Education,* **46,** (1), 113–127

Woolf, K., Potts, H.W.W. & McManus, I.C. (2011) 'Ethnicity and academic performance in UK-trained doctors and medical students: systematic review and meta-analysis', *British Medical Journal,* (342). DOI https://doi.org/10.1136/bmj.d901

Wyness, G. (2017) *Rules of the game: disadvantaged students and the university admissions process.* London: The Sutton Trust

SECTION 3:

Science teachers: synthesising learning

CHAPTER 11

Creativity in teaching science

Deb McGregor & Nicoleta Gaciu

This chapter discusses the meaning of creativity and what it looks like in the classroom. It also provides many suggestions for how teachers can alter their practice in small ways to be creative themselves, as well as encouraging creativity in their students, including approaching everyday objects and events in an unusual way, presenting intriguing phenomena, asking questions and supporting students to think creatively.

What is creativity?

Creativity can be understood and thought about in various ways. If we ask children or adults about it, often they will describe a view that relates it to the Arts. If we ask them to define it, they may say that it is being able to *play* a musical instrument, *paint* a picture, *act* a part in a play, or *write* a unique song, poem or story. This indicates how people generally understand creativity as a 'performance-related' capability. If asked *who* is creative, they may reply 'Van Gogh', an impressionist painter, or 'Shakespeare', a playwright. Asked about scientists, they may suggest Albert Einstein or Isaac Newton, both of whom developed novel theories about forces and the motion of bodies in space. This kind of 'Big C' creativity of global eminence (Kauffman & Beghetto, 2009) is unlikely to be evidenced in school science classrooms. However, if we ask teachers about students' creativity in their lessons (Andiliou & Murphy, 2010), they notice that they can be innovative, imaginative and inventive in various ways. They can apply newly acquired strategies to novel tasks, and the process of arguing, decision-making, evaluating and assessing ideas, through critical review, can inform creative alternatives or fresh perspectives.

Student creativity is more akin to 'little c', a novel idea that is fresh to them, the originator(s), or 'mini c', that is everyday creativity (Craft, 2001, 2010). Individuals can, therefore, be creative in varying degrees, ranging from those (like Einstein and Newton) who demonstrate profound and significant originality ('Big C'), to others, such as students in school, who may develop original ideas in their own terms ('little c'), illustrated in the conversation below. Within the topic *materials and their properties*, these students in Wellington, New Zealand, are discussing the implications of playing sport (netball) in a world made of alternative materials (in this case, sponge):

Student 1: *'If the ball was made out of sponge, you would definitely need to, umm, throw harder because it would be like throwing a leaf, you...it wouldn't move that far.'*

Student 2: *'Yeah, but if you like caught the ball and stood still because like you can't move when you've got the ball, you'd start sinking.'*

Student 3: *'Yeah, imagine the court being sponge. If it rained you would like, sink.'*

Student 1: *'So you pivot, you jump, you throw, you pass, you shoot, that kind of stuff and then, if it's sponge, then you'll flop around, trip over.'*

Student 2: *'Sinking like quicksand, you would have to wear snow shoes, because snow shoes, like, spread your weight.'*

These students recognise the implications of increased air resistance when trying to throw the ball, and how the lack of friction and the pull of gravity would affect how they would move on the spongy surface. They suggest an inventive remedy, snowshoes, which will increase the surface area that comes into contact with the spongy surface (and should improve the stability and rapidity of their movements). Making creative connections and directly linking them to daily life is also evidenced by another group from a school in Derby, UK. At the end of a lesson on turning moments, which has involved reviewing the use of a wheelbarrow to remove garden waste, students are asked to reflect on the application of the relationship between 'lift' and 'load'. A clear illustration of 'mini c' is offered when they come up with the bright idea that wheelbarrows could have extendable handles so that the distance from the fulcrum (the wheel) is increased, which would facilitate heavier loads being more easily moved in the garden.

At the other end of the scale, 'Big C' ideas include Carson's recognition of the effect of pesticides on the gradual population decline of songbirds; Galvani's discovery of electrical impulses in frogs' legs; Lavoisier's observations relating non-visible gases and the mass of solid substances, and Jenner's inspiring hypothesis (deemed somewhat unethical today) that 'giving' someone a disease can prevent another similar, but more deadly, one developing! These kinds of inventive ideas emerged by looking at and thinking about observations from fresh perspectives (such as Newton keenly theorising how bodies move), or reviewing the character and behaviour of materials and objects (such as Mendeleev systematically categorising similar substances to generate the first Periodic Table), or thinking about how to apply principles from the natural world (such as Georges de Mestral recognising how the structure of burrs could inform the generation of a new material like Velcro). Each of these innovative thinkers has contributed to our understanding about the nature of the world in which we live today. So perhaps, by adopting more creative teaching strategies (Kind & Kind, 2007 p.5) rather than continuing to only use traditional approaches (see Table 1), we may nurture more inventive thinkers for the future.

Table 1: Pedagogic contrasts of creative and more traditional teaching (after Kind & Kind, 2007)

Creative teaching	Traditional teaching	
Student-oriented	Teacher-oriented	
Group/team work	Individual work	
Co-operative learning	Individual learning	
Explorative tasks	Close-end tasks	
Open-ended problems	Closed problem	
Open investigations	'Recipe' work	
Hands-on teaching	Lectures	
Outdoor activities	Classroom activities	
Project work	Lectures	
Issue-oriented	Concept-oriented	
Teachers taking risks	Teachers playing safe	

Creativity in education

Guildford (1950) first articulated the debate about creativity in education through his recognition of the difference between convergent and divergent thinking. This was drawn upon by the National Advisory Committee on Creative and Cultural Education (NACCCE) in 1999, which highlighted concerns about the capacity of future generations to be creative enough to deal with any unforeseen scientific or societal problems that might arise in the new millennium. The NACCCE identified four characteristics of creativity: *Imagination* and *Purpose; Originality* and *Value* (1999, p.13) and described teaching creatively as *'using imaginative approaches to make learning more interesting and effective'*. QCA (2004) extended these teaching features to consider further what learners could be doing to illustrate creativity. They identified:

➤ Questioning and challenging conventions and assumptions;

➤ Making inventive connections and associating things that are not usually related;

> ➤ Envisaging what might be, through imagining and seeing things in the mind's eye;
> ➤ Trying alternatives and fresh approaches by keeping options open; and
> ➤ Reflecting critically on ideas, actions and outcomes.

They also highlighted that *'When pupils are writing a poem, choreographing a dance or producing a painting, their work can be unique if it expresses their ideas and feelings. But what about work in subjects like science, history and maths? While it would be wonderful for a pupil to be the first person to discover a new scientific principle, this is highly unlikely. Does this mean that pupils can't be creative in these subjects? Not at all. Skilled teachers can help pupils tackle questions, solve problems and have ideas that are new to them. This makes pupils' ideas original, the result of genuinely creative behaviour'* (QCA, 2004 pps.77–78).

QCA subsequently advised that *'...with minimal changes in planning and practice teaching...teachers can promote children's creativity'* (p.4). Jeffrey and Craft (2004, p.1) clarified how Teaching Creativity (TC), or innovatively *differed* to Teaching For Creativity (T4C), intended to *'develop young people's own creative thinking or behaviour'*.

More recently, Ofsted (2010) has even paid attention to creativity that can *raise standards* and suggests that teachers should be *'...encouraging pupils to question and challenge, make connections and see relationships, speculate, keep options open while pursuing a line of enquiry, and reflect critically on ideas, actions and results'* (Ofsted, 2010 pps.5–6).

This chapter, consequently, suggests a range of ways in which practice (TC *or* T4C) could be adapted or adopted (see Tables 2 and 3) to promote and develop creativity in science lessons.

What can creativity look like in a science classroom?

It is difficult to summarise exactly *how* a teacher can become more creative in the classroom, because a teacher's personal qualities and values, pedagogy and the school ethos may all combine to contribute to dimensions of creative practice. Teachers who value a creative environment may imaginatively decorate their classrooms with the intention of communicating, modelling and stimulating innovative thinking. This can include: creating dynamic displays on laboratory walls; generating an area that represents or illustrates an aspect of science, such as a miniature tropical rain forest ecosystem or suspended models (from the ceiling) of stars, planets and moons in our solar system; or an exhibition including everyday materials that comprise the elements of the Periodic Table. A teacher (or a visitor) may dress like Robert Hooke when teaching about cells, be the director of a play featuring Galvani and Volta arguing about the nature of electricity (Davies & McGregor, 2017 p.117), demonstrate an impressive series of 'wow' demonstrations (that produce stunning reactions), or even present experiments that appear to be magic (Spangler, 2017), or entice students to venture close to a Van de Graaff generator (Sang, 2016 p.181) to produce the spectacle of vertical suspension of the subject's hair. These are all creative ways to communicate about science. So, although science has traditionally been a secondary school experience that emphasises the straightforward presentation of key scientific ideas related to biology, chemistry and physics, there are generalised ways (as suggested in Tables 2 and 3) in which teachers can think afresh about how they explain, communicate and convey concepts to students more imaginatively and creatively. As Robinson (2001, p.111) suggests, *creativity is possible in any [human] activity*. Frodsham (2017) identifies how teachers can be creative in different ways, but she warns that innovative practice does not necessarily support creative learning.

Table 2: Dimensions of Creative Practice, adapted from Grainger *et al* (2005) and Davies and McGregor (2017)

Personal qualities	Pedagogy	School ethos
Commitment to children Desire to learn Flexibility and enthusiasm Risk-taking and curiosity Understanding children's needs and interests Using humour Secure knowledge base	Using diverse teaching methods Identifying entry points for individuals Linking ideas Connecting with students' lives Using (the latest and) current technology Adopting a questioning stance Encouraging students to ask questions Encouraging and valuing independent thinking Encouraging students working together to develop ideas and suggestions	Environment that values (useful) thinking and doing differently Environment promotes emotional engagement Students feel safe, valued and trusted Students encouraged to speculate and take risks Appropriate resources provided Leadership that supports and celebrates creativity Links with the wider community

Table 3: Suggestions for teaching creatively and teaching for creativity
(adapted from Davies and McGregor, 2017)

To teach creatively you could...	To teach for creativity you could...
• Review established activities (and practice) with creativity in mind • Make the ordinary fascinating • Share a sense of awe and wonder • Look at things differently • Maximise opportune moments: (i) reflecting on the unexpected; (ii) relating to events • Connect and relate science to everyday life • Actively illustrate or visualise scientific concepts and processes • Value questions • Encourage autonomy in investigations	• Consider ways that unexpected creativity can be valued as a Learning Outcome • Encourage (and value) students' questions (re: awe and wonder – remove when done) • Encourage agency (more autonomy) • Encourage 'experimenting' with project/practical materials • Generate authentic inquiry opportunities • Celebrate plausible creative ideas or diverse thinking • Celebrate thoughtful actions (that are not teacher-directed) • Recognise and value students' creative or alternate interpretations of scientific evidence/observations • Create 'space' for philosophic discussions about science • Ensure that students appreciate and understand what creativity/innovation and originality look like in a science classroom • Include valuing creative ideas or innovative suggestions in peer-peer assessments

Review established activities (and practice) with creativity in mind

Teaching about floating and sinking (Archimedes' Principle) tends to be taught through students being directed to place different objects, in turn, in a tank of water to see whether they float or sink. Usually a systematic table of results is produced indicating whether objects such as a cork, metal block, plastic bottle, etc. sink or float. A more creative approach could be to adopt (and twist a little) the story of Archimedes and why he exclaims *'Eureka'*, to generate more of a puzzle. The teacher could describe how *'Archimedes was playing with a small boat in a bath. When the boat was filled with stones, it was still floating, but when it accidentally tipped over, the stones fell to the bottom of the bath (without letting any water into the boat)'*. At this point, the imaginative teacher could say *'Did the water level go up, stay the same, or go down?'* (Davies & McGregor, 2017). The teacher may not know the answer, but it remains an appropriate question to ask because the students can think about how to find out what happens and try it out. Not being able to actually use a real bath to do this in the science laboratory, though, means that the students will have to come up with a variety of approaches to solve the problem. Allowing them time to discuss ways in which they might do this, with measuring cylinders, beakers, jugs of water, a model boat and assorted pebbles, a balance, etc., can facilitate creative thinking. However, as a teacher, not knowing what the answer might be requires a risk-taking attitude (as identified in Table 2). Creative teachers are flexible and enthusiastic, and taking a risk might mean sharing that they don't know *all* the answers, but this in turn can give students greater confidence to express their own ideas.

Studying the human body, for example, is often taught through students drawing, labelling and annotating diagrams of various systems, organs or tissues. This is a relatively passive form of learning, so using games like 'Locate an organ' or 'Bolus beetle drive' (Lock, 1991b, 1998), or adaptations of TV shows such as *Blockbusters*, *Countdown* or *Pointless*, can all offer a more interactive and enjoyable way to engage with biological (or any other science) subject matter. Lock (1991a) also provides tried and tested illustrations of ways that expressive and poetic writing, cartoons, comics and posters can be produced. Each of these alternate approaches can be applied to any aspect of the science curriculum. These creative forms of written expression engage and involve students in thinking more precisely and accurately (Lock, 1991a p.39) about science. To encourage more practical skills in science, competitions can be introduced (Longshaw, 2009): for example, one that could be used to introduce the Bernoulli principle. Given materials including a straw, ping pong ball, scissors and a two-litre plastic soda bottle (Spangler, 2017), students could be asked who can levitate the ball for the longest or the highest above the end of the straw. Connecting practical challenges, such as *Who can lift an ice cube with string?* (Planet Science, 2010), with teaching changes of state, melting points and so on can inject a fun opportunity to develop application of scientific knowledge. The Planet Science website is a rich source of science experiments, tricks, stories and even jokes that can be adopted and adapted by creative teachers to introduce or integrate into any topics of the science curriculum.

Making the ordinary fascinating

Many young people do not realise that science is all around them, and the formality of school education often means that the fascination of the everyday world around them is overlooked. Making breakfast at the beginning of the day may involve changes of state if eggs are boiled, fried or poached and viscous porridge is made from dry oats. Everyday happenings, however, become mundane and students often forget to 'look' or 'think about' science at home or on the sports field. Conveying a sense of fascination in the ordinary requires the resourceful teacher to make the appropriate connections.

Collier (2006) proposes that a creative way to stimulate students to think about the optical properties of different materials can be promoted with a display of various items, including paperweights, magnifying lenses, tumblers, mirrors, opaque glass, coloured glass and objects with fibre optics. Altering the direction of lighting or even introducing colours to illuminate the display can provide additional visual effects to further pique interest.

As Davies and McGregor (2017) describe, soil seems less promising, yet even a scoop of this can be fascinating if stirred into water in a large, clear jar and allowed to settle over the space of a few hours. 'Magically', the variety and colour of different grain sizes are revealed as the soil sinks to form graded beds. Liquids of different densities too can create a very visual impact. Everyday liquids such as oil and water will separate to form distinct layers. However, putting vegetable oil, water (with food colouring added) and syrup into a jar will quite quickly produce three differently coloured layers that intrigue observers. McGregor and Gunter (2006) adopted this phenomenon to generate creative thinking and invite predictions from students to suggest how liquids found in a kitchen cupboard might behave when mixed together. Some students are surprised by cooking oil floating on water; others by syrup sinking below the water. Inviting students to subsequently predict where in the 3-layered liquid column food items might settle provides rich opportunities for conjecture by drawing on observations in everyday life to try and accurately predict (and justify) how peanuts, grapes, tomatoes, rice grains and differently-shaped pasta might behave when dropped into familiar liquids. The rather colourful layers within which some objects float (a fresh cherry tomato, for example, will float on top of the water layer because of the air spaces in its tissue) and others sink (such as the peanut …eventually, depending on the amount of oil in it and the amount of salt on its surface) engage learners in thinking quite differently about density, mass and volume (and surface tension).

Lock (1991b) also offers suggestions for the ways in which ordinary, everyday materials, such as plastic bottles, cardboard boxes, string, etc., can be used to create original 3D models of scientific entities, such as the digestive system, the structure of different teeth, a cross-section of a leaf, the differences between plant and animal cells, even protein synthesis or a part of the DNA helix.

Sharing a sense of awe and wonder

Sharing an amazing observation about the world around us with students can be really infectious. Using unfamiliar objects and the story of the discovery of light, an Oxfordshire teacher uses artefacts to hold the students completely spellbound. She has a chronologically organised display, including pieces of rock, flint, moss, candles made of animal fat and different waxes, 18th Century candle-cutting scissors, a collapsible 19th Century lantern, an early 20th Century paraffin lamp and even a rotating Christmas tree decoration powered by a candle. Many of these objects students have never seen before. She draws on a range of historical stories, she explains how fire was discovered and uses the various items to illustrate her narratives, including a 120 year-old tinderbox that contains a sulphur-tipped match and dried fibrous hemp. She spends around an hour detailing how Cheese-grater, Coach and Chinese lamps work differently to provide light for cave dwellers, Edwardians, Victorians and, eventually, modern homes today. The students are fascinated, listening intently (and watching her use the various objects) as she tells her stories. As well as demonstrating how interesting and useful historical artefacts are, teachers can also reveal awesome natural processes, such as spectra or rainbows that can be created in the classroom using strong light sources (with blackout) and prisms. Bubbles are also enthralling for younger students. There are ways to make bubbles within bubbles, square bubbles, bouncing bubbles, smoky bubbles and even multicoloured bubbles (see https://www.stevespanglerscience.com/lab/experiments/giant-bubble-experiment/). Visual phenomena such as these can truly provoke students' imaginations to wonder how they are made, and what they could do to produce some themselves!

Looking at things differently

We now know quite a lot about the intuitive (and often quite naïve) ways in which young people interpret the world around them (Driver, 1983). Students can be forgiven, for example, for initially thinking that a circuit only requires a power source and one lead to the electrical appliance in order to work, if they base their ideas on observations made at home, where they 'see' a TV or a computer with one lead plugged into an electrical socket at the wall. This is a little like the young girl who Frodsham (2017) describes, who sees a spider's web spanning the branches of a cotoneaster and, because there are red berries close by, believes that they are the food for the arachnid and that's why the spider has spun its web there. These young people have drawn on their own experiences and knowledge to make sense of their observations. Helping learners 'see' things from different perspectives could enable them to generate more accurate understandings about science. Davies and McGregor (2017) suggest that students could be shown optical illusions as a way of illustrating that what we first 'see' is not necessarily the same as other people, and that, over time, explanations from others (and the teacher) about what they can see can facilitate the brain decoding the visual image. Using a 'Magic Eye' book, it is possible for students to see whether they can 'de-focus' their eyes sufficiently to see the hidden 3D images camouflaged in the abstractly patterned pages. Suddenly 'seeing' the hidden image can be rather spectacular!

Scientists such as Leeuwenhoek developed the first microscope to observe what he called 'animalcules', including bacteria, sperm, protozoa and living cells. Robert Hooke went on to see cork cells. Using magnifiers can enable learners to see beyond what is possible with the naked eye, and illustrates how scientists are keen and careful observers. McGregor and Precious (2015, p.9) also suggest a number of ways to see differently. They describe a variety of ways in which a portable magnifier enables teachers to show previously unseen parts of everyday objects in dramatic ways. This kind of technique involves careful observations, especially if an object is magnified and viewed from intriguing angles, like peering down the water-filled trumpet, focusing on the lid, and then the hairy surface, of a carnivorous pitcher plant (see https://pstt.org.uk/resources/cpd-units/dramatic-science). Even if real objects or organisms are difficult to obtain, using enlarged photographs and pictures is another way to promote making the ordinary fascinating.

Another inventive approach that can be used to make visible those processes unseen with the naked eye was used by a teacher from Stoke-on-Trent who asked all her class to rub some hand cream into their hands. Then, just one student dipped her hands in glitter. The class was told to walk around the room, shaking hands with each other as if they were old friends meeting up. At the end of 10 minutes, most of the students had glitter on their hands and several also round their faces and mouths (despite the single source of glitter!). This was an enjoyable way to illustrate how quickly bacteria can spread through touching hands, and also highlighting how, without thinking, someone infected may touch his/her face or other objects in the room. This can be repeated after washing hands, to illustrate the importance of hygienic processes in preventing the spread of disease.

Sherborne (2017) provided a section on the *upd8.org.uk* website, which dealt with 'difficult and dull' aspects of the curriculum. These resources provided quite imaginative ways to engage students with scientific concepts including plants, waves, rocks, particles, electricity and magnetism, pressure and turning moments.

Maximising opportune moments

(i) Reflecting on the unexpected

'Much of the artistry in being a successful teacher involves holding on to the notion of possibility in what may seem to be adverse situations' (Craft, 2000 p.3). When an inquiry or investigation 'goes wrong' or something unexpected happens, it is tempting to assume that the equipment is faulty and ignore the findings. However, if a series circuit that includes a lamp and buzzer results in a beeping sound but no light, a creative teacher of science would use such occurrences to suggest new ideas for inquiry rather than just 'give up' and suggest that alternative equipment should be used. An inventive teacher could say: *'Wow! That's interesting. Do you think any other combinations of two different components in a circuit will do the same thing? What if you put a buzzer and motor together? Has anyone got any other ideas or suggestions?'.* Hopefully, this would spark student ingenuity to suggest alternate strategies, and even offer a novel explanation of the surprising result (what is likely to have happened is the high internal resistance of the buzzer reduces the current all around the circuit to such a low level that, although it flows through the lamp, it doesn't actually make the filament hot enough to glow). McMahon (2006) suggests that changing weather and seasons can also provide opportunities to look at habitats in different conditions; a dewy autumn morning might be a good time to go spotting spiders' webs. When the early morning dew or puddles disappear in the school grounds, because they evaporate, it might be useful to consider where else in the environment water is transformed from a liquid to a gas. After a fall of snow overnight, the way in which humans keep warm and the thermal insulation properties of cold weather clothing can be considered. There is also the opportunity to invite students to consider why parents always insist that, in winter, wet clothing should be removed and replaced with dry clothing (this could be investigated to explore how wet/ dry clothing of different types promote or prevent failing body temperatures, and provoke discussion about the potential onset of shivering or even risk of hypothermia).

In many secondary schools, Cognitive Acceleration in Science Education (CASE) (Adey, Shayer & Yates, 2001) activities set up situations where students are purposely offered the opportunity to explain unexpected events or observations. An example of this approach to present something unanticipated to promote students' creative explanatory thinking is the use of three candles burning in three different-sized beakers. Repeats of this demonstration can provide 'evidence' (McGregor, 2004) that the candle in the middle-sized beaker burns the longest. This somewhat perplexing observation is an invitation for students to develop plausible explanations involving the role of oxygen in combustion, the production of carbon dioxide and convection currents that are generated in the upside down beaker to affect the burning times of the candle wick. Burning candles at Christmas or in the summer (to ward off insects) can be linked topically to the time of year.

(ii) Relating to events

Using topical news items as a stimulus for science work is innovative and assists students in feeling that their lessons have relevance. Linking to the news through *Newsround*, the BBC television programme aimed specifically at younger people (see www.bbc.co.uk/newsround/) provides innovative resources and ideas. The *Upd8* (www.stem.org.uk/resources/collection/2827/science-upd8) and *Engage* (www.engagingscience.eu/en/overview/) websites also provide many useful links to societal concerns such as global warming and the use of diesel. These websites develop current news stories and turn them into science lessons, a useful way to keep your practice original and up-to-date.

Teaching about forces, including gravity, is related to the Egg Drop Challenge (see www.sublimescience.com/free-science-experiments/egg-drop-challenge/). This can be associated with a need to deliver food from helicopters to people of the Caribbean islands after the recent hurricanes. Some foods (such as eggs) are delicate and survivors need food that can be dropped from helicopters, providing the contextual justification for the design and testing of the best egg carrier.

Also related to natural disasters, several lessons were presented by a creative teacher from Sheffield who adopted a desert island shipwreck theme, where the whole class created a huge desert shelter on the school field. She thought that this was innovative because they had very little material provided and had to think/work together to turn natural materials into a sturdy shelter!

At Christmas time, there are always strategies to bring tricks into the classroom to pique students' curiosity and creative speculation. The Steve Spangler website (see www.stevespanglerscience.com/lab/experiments/) is a rich resource for many themed experiments (including Halloween and Easter too). Lock (1992) creatively offers Yuletide scientific problems to solve, which could be adapted or adopted as investigations. The Institute of Physics (IOP) also offers some quite spectacular practicals for special events such as open days (see www.practicalphysics.org/files/Paul_Gluck_PED_articleSep07.pdf).

Connecting and relating science to everyday life

Students do think of science as somewhat elitist, so relating science activities directly to themselves can promote more interest. Thinking about protecting their own skin when outdoors can be effective using UV bracelets (available from www.stevespanglerscience.com/store/uv-color-changing-beads.html). One teacher from North London reported: *'I gave them all a "special science bracelet", asked them to wear it all day and then try and explain what made it special. It was made of UV beads that change from white to bright colours when in UV light. Most thought it was temperature that changed the beads because it was a cold, cloudy day and they changed colour at break when they were outside. I asked them to take them home and test them there too. Many came back the next day saying it was light that made them change because they had been outside in the cold (but dark) evening and the beads hadn't changed colour. Following this, I asked them to predict what would happen if we sprayed the beads with sun cream. After we did this, we discussed the similarities between these beads and our skin; now they understand why they need to wear sunscreen! Six months on, many children are still wearing their special science bracelet every day!'* (Frodsham, 2017).

Using stories of scientists' lives (McGregor & Precious, 2015 pps.171–227) can also help students appreciate how these are ordinary people who have carried out their work and pursued lines of inquiry before their revelations or new theories emerge to help us better understand the world around us. There is a BBC website (www.bbc.co.uk/programmes/b06vy2jd/episodes/downloads) that offers podcasts with some useful and surprising stories about scientists and their discoveries.

Actively illustrate or visualise scientific concepts and processes

Students sometimes find the rather abstract ideas in science difficult to understand. Acting out processes or concepts requires clarity in understanding to be able to represent the science accurately. A teacher at King's School, Auckland (reported in Davies and McGregor, 2017) guided her young children to explore the relationship between the pitch (frequency) of a sound and the length of vibrating object, by encouraging them to bring in their musical instruments from home to demonstrate how they could make different notes. After looking at how a trumpet, saxophone and flute could change the length of the column of vibrating air, she challenged the class to make a paper straw 'clarinet', by cutting one end into a 'v', flattening it and blowing through. It took a bit of practice for some students to get the lip position and force of breath right, but they all eventually managed to produce a 'toot', a little like the sound some people can make by blowing over a blade of grass held between thumbs. Making their 'clarinets' different lengths, by cutting pieces off the other end of the straw as they blew, very clearly demonstrated pitch rising as the length shortened. The teacher then challenged the students to make more than one note from the same straw, which produced some ingenious solutions such as changing how hard you blow, compressing the lips, adding a sliding section like on a trombone, or holes as in a recorder. Other examples of performing scientific concepts are clearly demonstrated by the ways in which Abraham and Braund (2012) use drama. They include a comprehensive range of biology, chemistry and physics enactments, ranging from adaption and survival to the pH scale, polymerization, force and acceleration.

Valuing questions

As science teachers, we recognise the value of students raising questions, but often we are hard pressed to cover the curriculum content in the time available, so there is little 'space' to allow for individual learners to ponder and consider shaping appropriate questions. As a result of our pressurised time in the classroom, investigational questions are usually pre-determined, not least because the resources required often need to be pre-ordered at least a week in advance! It is therefore challenging to allow students to generate their own questions to be investigated. In primary schools, creative teachers have a 'questions board' and some may even develop a 'wonder wall' to which the teacher and students can contribute as conundrums or unanswered queries arise; this could be introduced into secondary classrooms. Reviewing these on a daily or weekly basis can become a celebration of creative thinking, as peers could consider which are the most thoughtful questions.

Encouraging autonomy in investigations

Investigating in science requires scientific ideas to solve problems (Roberts, 2009). In attempting to respond to an hypothesis or an open-ended query of some kind by planning and carrying out an original investigation, students doing school science are to some extent *'mimicking real scientists' creativity'* (Kind & Kind, 2007, p.9). This aspect of creative practice is arguably the one with which most secondary science teachers are comfortable, as 'working scientifically' (DfE, 2014) resonates with science degree experiences. It can offer students the opportunity to show 'imaginativeness' in their scientific ideas or ingenuity in the way that they approach an investigation, or solve practical problems that they encounter as they progress through their inquiry. Roberts (2009) suggests that encouraging students to think about evidence in particular ways can promote creativity. She also offers evidence to suggest that, without *both* substantive factual knowledge *and* procedural understanding (*ibid*, p.31), creativity in open-ended investigations may be curtailed. She suggests how interventional teaching that emphasises consideration of alternate possible solutions to scientific problems can increase imaginative engagement with scientific ideas.

Ofsted (2008) highlighted how *stimulating teaching* and *enthusiastic learning* (where TC and T4C combine) encouraged the development of inquiry skills: '...*the most stimulating teaching and most enthusiastic learning occur when teachers encourage their pupils to come up with ideas and suggestions and, in consultation with their teacher, to plan, conduct, record and evaluate their own investigations. Good formative assessment is also crucial to success. When pupils receive regular feedback on how well they are progressing and clear advice on how they can improve further, they are able to focus their energies effectively*' (Ofsted, 2008 p.4).

Moments in inquiry situations or investigational tasks during which students can engage in discussion and decision-making inherently involve exchange and development of creative suggestions, alternate ideas and approaches. These can be encouraged by the ways that teachers *ask open questions, invite collective suggestions* or indeed *provide time for groups* of students to *work independently* on answering their own scientific question. What is important is to invite and *celebrate plausible and reasoned inventive or imaginative ideas* because, as Perkins (2000, p.3) reminds us, '*Leonardo da Vinci was wrong. But he was insightfully wrong. He came to a mistaken idea about flight, but the pattern of thinking behind the idea was exemplary*'. However, as we know, da Vinci made keen scientific observations (of human anatomy) and developed mechanical engineering designs that drew on scientific principles. So, as teachers, we should encourage creativity because, although in our classrooms students may not produce the next big idea in science, they may become scientists in their adult working lives, and certainly may be creative enough to solve their own everyday problems by applying understandings and approaches that they have experienced and developed for themselves in school science.

Challenges to developing creativity in science lessons

Although teachers generally recognised that they could nurture creativity in their classrooms, a significant number (Andiliou & Murphy, 2010) indicated that school climate, curricular guidance and testing procedures constrained the extent to which they felt they could use lesson time to support and encourage their students to be creative. It seemed that teachers thought that acquiring a large body of knowledge was a hindrance and incompatible with developing creativity. Roberts (2009) also argues that, for teachers, it is challenging to develop creativity in science because it is not readily available as a usable recipe in a textbook. Andiliou and Murphy (2010) also note how there is little professional development available for teachers to appreciate how, when and where they can most effectively develop not only their own creativity in teaching, but also support that of their students too.

In conclusion

Creativity is important in secondary science education and, although advice about how to nurture it in classrooms is still developing, adopting the approaches and strategies outlined in Tables 2 and 3 can certainly contribute. Creativity is acknowledged globally as a skill or aptitude that will be influential in helping resolve unknown economic, scientific and technological problems that we might face in the future. So, for both individual benefit and societal advantage, creativity should remain prominent and visible within the school science curriculum.

References

Abrahams, I. & Braund, M. (2012) *Performing Science. Teaching Chemistry, Physics and Biology through Drama.* London: Continuum

Adey, P., Shayer, M. & Yates, C. (2001) *Thinking Science.* London: Nelson Thornes

Andiliou, A. & Murphy, P.K. (2010) 'Examining variations among researchers' and teachers' conceptualizations of creativity: A review and synthesis of contemporary research', *Educational Research Review,* **5,** (3), 201–219

Collier, C. (2006) 'Creativity in materials and their properties'. In *Creative Teaching: Science in the Early Years and Primary Classroom,* Oliver, A. (Ed.). London: David Fulton

Craft, A. (2001) 'Little c Creativity'. In *Creativity in Education,* Craft, A., Jeffrey, B. & Leibling, M. (Eds.), pps. 45–61. London: Continuum

Davies, D. & McGregor, D. (2017) *Teaching Science Creatively: Learning to Teach in the Primary School Series (2nd Edition).* Abingdon: Routledge

Department for Education (2014) *National Curriculum.* Available at: www.gov.uk/national-curriculum Accessed 28.09.17

Driver, R. (1983) *The Pupil as Scientist?* Milton Keynes: Open University Press

Frodsham, S. (2017) *Developing creativity within primary science teaching. What does it look like and how can classroom interactions augment the process?* Unpublished PhD thesis. Oxford Brookes University

Grainger, T., Goouch, K. & Lambirth, A. (2005) *Creativity and Writing: Developing voice and verve in the classroom.* London: Routledge

Guilford, J.P. (1950) 'Creativity', *American Psychologist,* **5,** (9), 443–443

Jeffrey, B. & Craft, A. (2004) 'Teaching creatively and teaching for creativity: distinctions and relationships', *Educational Studies,* **30,** (1), 77–87

Kauffman, J.C. & Beghetto, R.A. (2009) 'Beyond Big and Little: The Four C Model of Creativity', *Review of General Psychology,* **13,** (1), 1–12

Kind, P.M. & Kind, V. (2007) 'Creativity in Science Education: perspectives and challenge for developing school science', *Studies in Science Education,* **43,** (1), 1– 37

Lock, R. (1991a) 'Creative work in biology – a pot pourri of examples. Part 1', *School Science Review,* **72,** (260), 39–46

Lock, R. (1991b) 'Creative work in biology – a pot pourri of examples. Part 2: Drawing, drama, games, models', *School Science Review,* **72,** (261), 57–64

Lock, R. (1998) 'Digestion aids? Games with guts', *School Science Review,* **80,** (291), 88–93

Longshaw, S. (2009) 'Creativity in science teaching', *School Science Review,* **90,** (332), 91–94

McGregor, D. & Gunter, B. (2006) 'Invigorating pedagogic change: Suggestions from findings of the development of secondary science teachers' practice and cognitive learning process', *European Journal of Teacher Education,* **29,** (1), 23–48

McGregor, D. & Precious, W. (2015) *Dramatic Science.* Abingdon: Routledge

McMahon, K. (2006) 'Creativity in life processes and living things'. In *Creative Teaching: Science in the Early Years and Primary Classroom,* Oliver, A. (Ed.). London: David Fulton

NACCCE (1999) *National Advisory Committee on Creative and Cultural Education: All Our Futures: Creativity, Culture and Education.* Available at: http://www.tracscotland.org/sites/default/files/Creativity%20Culture%20and %20Education%20intro%20and%20summ.pdf Accessed 28.09.17

Ofsted (2008) *Success in Science.* Available at: http://webarchive.nationalarchives.gov.uk/20141107070713/http://www.ofste d.gov.uk/resources/success-science Accessed 28.09.17

Ofsted (2010) *Learning: creative approaches that raise standards.* Available at: http://webarchive.nationalarchives.gov.uk/20141116062720/http://www.ofste d.gov.uk/sites/default/files/documents/surveys-and-good- practice/l/Learning%20creative%20approaches%20that%20raise%20standar ds.pdf Accessed 28.09.17

Perkins, D. (2000) *The Eureka Effect. The art and logic of breakthrough thinking.* New York: Norton

Planet Science (2010) *Who can lift an ice cube with string?* Available at: http://www.planet-science.com/categories/experiments/magic-tricks/2011/12/can-you-pick-up-ice-with-a-piece-of-string.aspx Accessed 01.10.17

QCA (2004) *Creativity: find it, promote it.* Retrieved from: http://www.teachfind.com/qcda/creativity-find-it-promote-it Accessed 28.09.17

Roberts, R. (2009) 'Can teaching about evidence encourage a creative approach in open-ended investigations?', *School Science Review,* **90,** (332), 31–38

Robinson, K. (2001) *Out of Our Minds: Learning to be Creative.* Chichester: Capstone Publishing Ltd.

Sang, D. (2016) *Teaching Secondary Physics.* London & Hatfield: Hodder Education and The Association for Scence Education

Times Educational Supplement (2017) *Harry Potter Experiments.* Available at: https://www.tes.com/teaching-resource/harry-potter-potions-activities-and-experiments-11427732 Accessed 28.09.17

Sherborne, T. (2017) *Engage Science.* Retrieved from: upd8.org.uk Accessed 01.10.17

Spangler, S. (2017) *Steve Spangler Science.* Available at: https://www.stevespanglerscience.com/lab/ Accessed 01.10.17

CHAPTER 12

Thinking about practical work

Ian Abrahams & Nikolaos Fotou

Most planning regarding practical work relates to getting students to do and observe things with objects and materials, rather than how students are to think about and learn from their actions and observations. This chapter presents an approach to thinking about practical work to help focus on students' learning in practical work.

Background

This chapter considers the effectiveness of practical activities in school science and suggests the need for a greater emphasis on a 'minds-on' approach to practical work. This fits with recent advice from the Gatsby Foundation on Good Practical Science, which gives ten benchmarks for improving practical sessions in schools (see www.gatsby.org).

It has been suggested (Donnelly, 1998) that a reason for the frequent and widespread use of practical work in secondary school science is that many science teachers see its regular use as an essential part of what it *means* to be 'a science teacher'. That practical work *'seems the "natural" and "right" thing to do'* (Millar, 2002 p.53) means that many teachers see its use as the basic *modus operandi* for the teaching of science.

The risk inherent in this approach is that its use can become so much a matter of routine that teachers cease to assess critically whether it is always the most appropriate way of achieving specific learning outcomes. Indeed, many teachers, if asked, say that they believe that practical work leads to more effective learning – students are, they feel, more likely to understand and remember things they have done, as opposed to things that they have just been told, or have read about, as the following examples illustrate (all names are pseudonyms):

Mr Oldstead: *'I think I believe strongly that by doing things you're more likely to remember it. I mean, if I look at my own kids they're more likely to remember stuff and be able to do things if they have a go at it, they're practising and I believe the kids here are the same.'*

Mr Saltmarsh: *'I think if things have gone well, in a specific practical, it does help the children to understand and remember what they've done rather than just writing it down.'*

Yet, despite this, we know both from experience and research (Abrahams & Millar, 2008) that, frequently, students do not learn from a practical task the things we wanted them to learn. Furthermore, in the medium to long term, many students can only recall specific surface details of the practical task they undertook and are unable to say what they learned from it, or the reason that they undertook it:

Researcher: *'Do you remember any practical that you did longer ago?'*
Student SH5: *'Yeah, we got like different chemicals in the tubes like blue liquids and then put like a red in with them and see what they turned out like.'*
Student SH6: *'Yeah, you mix a and b, like copper sulphate and something else, and you mix it like together.'*
Researcher: *'And that was to help you learn what?'*
Student SH6: *'I don't know really.'* [Both students are laughing loudly.]
Student SH5: [Shrugs shoulders and shakes head to indicate that he also does not know.]

These issues have led some science educators to question the contribution of practical work to learning. For example, Hodson (1991) claims that *'as practised in many countries, it is ill-conceived, confused and unproductive'* (p.176), whilst Osborne (1998) suggests that practical work *'only has a strictly limited role to play in learning science and that much of it is of little educational value'* (p.156). Perhaps the key phrase in Hodson's claim is 'as practised'. Whilst few would doubt that practical work is an essential part of science education, the question we need to consider is whether we use it effectively. In order to answer that question, we need first to consider what we mean by 'effectiveness'.

Effectiveness

When thinking about the effectiveness of a teaching/learning task of any kind it is useful to consider the steps in both developing such an activity and monitoring what happens when it is used. To do this, a model of the processes involved in designing and evaluating a practical task, developed by Millar *et al* (1999, p.37), has been used and this is shown below in Figure 1.

Given that the aim of this model is to consider the effectiveness of a specific task *relative* to the aims and intentions of the teacher, the starting point (Box A in Figure 1) is an evaluation of the teacher's learning objectives in terms of what it is they want the students to learn. Once the teacher has decided

Figure 1: A model of the process of design and evaluation of a practical task
(From Millar *et al.*, 1999, p.37)

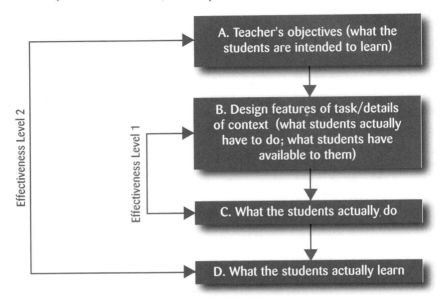

what they want the students to learn, the next stage (Box B) is for him/her to design a specific practical task (or use an existing one from a scheme of work) that s/he considers has the potential to enable the students to achieve the desired learning objectives.

However, because the students might not do exactly as the teacher intended when the task was designed, the next stage in the model in Figure 1 (Box C) considers what it is that the students actually do as they undertake the task. There are various reasons as to why, and to what extent, what the students actually do might differ from what was intended by the teacher. Students might, for example, not understand the instructions or, even when they do and adhere to them meticulously, faulty apparatus might prevent them from doing what the teacher intended. Alternatively, even if the task is carried out as intended by the teacher and all the apparatus functions as intended, the students still might not engage mentally with the task using the ideas that the teacher had intended them to use. The last stage in the model (Box D) is therefore concerned with the question of what it is that the students actually learn as a consequence of undertaking the task. The use of this theoretical model allows the question of the effectiveness of a specific practical task to be considered at two separate levels.

The first level of effectiveness relates to the issue of what students *do* relative to what the teacher intended them to do. This level of effectiveness, referred to as 'level 1 effectiveness', is about the relationship between boxes B and C in the above model. The second level of effectiveness considers what the students *learn* relative to what the teacher intended them to learn. This second level of effectiveness, referred to as 'level 2 effectiveness', is about the relationship between boxes A and D in the model. This model can therefore be used to clarify what is meant by the 'effectiveness' of a specific practical in terms of:

> ➤ Does the task enable the students to do the things the teacher actually wanted them to do when they chose to use that specific practical task?
> ➤ Does the task enable the students to learn what the teacher actually wanted them to learn when they chose to use that specific practical task?

By combining this model of effectiveness with a model of knowledge (Tiberghien, 2000), in which there are two distinct domains: the domain of observable objects and events (o) and the domain of ideas (i), it is then possible to consider each of the two levels of effectiveness in terms of these two distinct domains. The effectiveness of any practical task can now be analysed and discussed in terms of two levels, with each level being further divided into two domains. In terms of task effectiveness, these levels are defined in the following way:

> ➤ A task is effective at level 1:o if the students *do* with the objects and/or materials the things that the teacher intended them to do and, as a consequence, they see the intended outcome.
> ➤ A task is effective at level 1:i if the students *think* about the task using the ideas that the teacher intended them to use.

At levels 2:o and 2:i, the issue of effectiveness relates to whether or not the task enables the students to *learn* the things intended by the teacher:

> ➤ At level 2:o a task is effective if the students *learn* and can recollect details about the objects/materials/events that they have observed and/or handled.
> ➤ At level 2:i a task is effective if the students *learn* and can recollect the scientific ideas that provide an explanation about the objects/materials/events that they have observed and/or handled.

These two levels of effectiveness, each of which can be considered with respect to the two distinct domains of knowledge, can be represented (Figure 2) using a 2 x 2 effectiveness matrix:

Figure 2: A 2 x 2 effectiveness matrix

Intended outcomes...	...in the domain of observables (Domain o)	...in the domain of ideas (Domain i)
...at level 1 (what students do)	Set up the equipment and operate it in such a manner as to see what the teacher intended.	Think about the task using the ideas intended by the teacher.
...at level 2 (what students learn)	To set up and operate similar equipment. Discover patterns within their observations/ data.	To understand their observations /data by being able to link them, using the ideas intended by the teacher, with the correct scientific theory.

Effectiveness at level 2:i is therefore a necessary requirement if, as has been suggested (Millar *et al,* 1999; Solomon, 1988; Woolnough & Allsop, 1985), an important function of practical work is to provide a link between the domain of observable objects and/or events and the domain of ideas.

If, as appears likely, a task can only be effective at level 2:i if it were also effective at both levels 1:o and 1:i, task effectiveness across level 1 appears a necessary prerequisite if a successful link is to be created between the two distinct domains of knowledge.

Doing with objects and materials

Despite the fact that closed 'recipe'-style tasks are likely to be thought of as dull and demoralising (Arons, 1993) and are unlikely to be perceived as either meaningful or engaging (Wallace, 1996), many practical tasks in secondary school science are at, or close to, the closed 'recipe' end of the continuum (Abrahams & Millar, 2008). Some teachers, despite stating their preference for open investigations, see the use of such tasks as a necessary means of ensuring that most students, irrespective of their academic ability, are able to set up and produce a particular phenomenon and analyse the results within the relatively short period of time available in one lesson:

Mr. Normanby: *'Often the practicals are designed to be pupil friendly. You know, to make sure that within your double they'll see, at least most of them will, what you want.'*
Dr. Kepwick: *'I think they need to come in, be told how to do it, and get a result.'*

The fact that 'doing with objects and materials' is widespread and very successful (Abrahams & Millar, 2008) strongly suggests that teachers see the effective generation of a particular phenomenon, and/or set of results, by the majority of their students as being their main priority.

Doing with ideas

Practical tasks, as Millar *et al* (1999) have pointed out, *'do not [or should not] only involve observation and/or manipulation of objects and materials. They also involve the pupils in using, applying and perhaps extending their ideas'* (p.44). Whilst 'doing' with objects and materials is relatively self-explanatory, 'doing' with ideas is less obvious and needs some clarification. The theoretical 2 x 2 matrix representation of practical work discussed previously distinguishes in the horizontal dimension between *doing* and *learning* and in vertical dimension between *observables* and *ideas*. In this context, the two quadrants on the right-hand side of the matrix refer only to ideas that, in contrast to observables, cannot be directly measured or observed. Doing with ideas refers to the process of 'thinking about' objects, materials and phenomena in terms of theoretical entities that are not themselves directly observable. Clearly not all thinking is synonymous with 'doing with ideas' – far from it. For example, a student can think about the readings on a voltmeter solely in terms of *observables* if they think of their readings *only* as being the numbers on a dial or scale. However, if they think about those voltmeter readings in terms of a non-observable property of batteries and other circuit components – i.e. of their being a measure of the voltage – as the following example illustrates, this is what constitutes 'doing with ideas':

Researcher: *'What type of circuit is this?'*
Student SK18: *'It's a series circuit.'*
Researcher: *'So what's the voltmeter measuring?'*
Student SK22: *'How much energy is going in and how much energy is coming out.'*
Researcher:*'And what will that tell you?'*
Student SK22: *'How much energy is lost.'*

Having clarified what 'doing with ideas' entails, it is important to remember that task effectiveness, in the context used here, is a measure of what students do with ideas *relative* to what the teacher intended them to do with them. In this respect, whilst the students can think about a task in any way that they wish, the task only has the potential to be effective (or ineffective) at level 2 if the teacher actually *intends* the students to think about the observables using specific ideas.

Linking observations and ideas

What has been found (Abrahams & Millar, 2008) is that there is a significant difference between the effectiveness of practical work in the domain of observables and that of the domain of ideas. Whilst students are frequently able to make the observations that the teacher intends, they rarely talk to each other, or to the teacher, using the ideas that underpin the observable features of the task and whose use would enable them to make scientific sense of their actions and observations.

Indeed, many teachers appear (tacitly or explicitly) to maintain an inductive 'discovery-based' view of learning – and to expect that the ideas that they intend students to learn will simply 'emerge' of their own accord from the observations or measurements, provided only that the students are able to produce them successfully. The underlying epistemological flaw in this viewpoint, and the practical problems to which it leads, have long been recognised (see, for example, Driver, 1975).

Since science involves an interplay between ideas and observation, an important role of practical work is to help students develop links between observations and ideas.

Figure 3: Practical work: Linking two domains of knowledge
(From Millar *et al*, 1999 p.40)

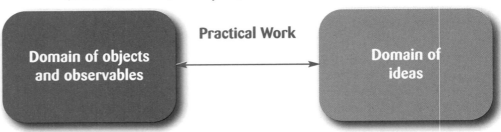

However, not only do these ideas have to be introduced, but it may be important that these ideas are used during the practical activity itself, rather than after it (possibly in a subsequent lesson) in order to account for what is being observed. Solomon (1999) discusses the valuable role that 'envisionment' can play in practical work, in terms of helping students to imagine what might be going on 'beneath the observable surface' as they do things with objects and materials and make their observations. Yet the evidence (Abrahams & Millar, 2008) is that few practical lessons are designed to stimulate an interplay between observations and ideas *during* the practical activity. Even if these links are developed in subsequent lessons, the fact that the ideas are not available at the time to help students make sense of the activity (to see its purpose), or their observations (to interpret these in the light of the theoretical framework of ideas and models), must arguably reduce the effectiveness of the practical task as a learning activity.

In terms of implications for practice, the suggestion here is that the two domains model, outlined in this chapter, is a useful tool for teachers when thinking about practical work. First, it draws attention to the two domains of knowledge involved, and their separateness – that one does not simply 'emerge' from the other. Second, it provides a means of assessing the 'learning demand' (Leach & Scott, 1995) of a particular task and of recognising that there is a substantial difference in the learning demand of tasks in which the primary aim is simply for students to see an event or phenomenon or become able to manipulate a piece of equipment, and those tasks where the aim is for students to develop an understanding of certain theoretical ideas or models that might account for what is observed. If teachers can be helped to distinguish more effectively between tasks of relatively low learning demand and those where the learning demand is much higher, this would then allow them to identify those tasks where students might require greater levels of support in order that the intended learning might occur.

To achieve this aim, it is necessary to ensure that task design more clearly reflects the fact that 'doing' things with objects, materials and phenomena will not lead to students 'learning' scientific ideas and concepts unless they are provided with a 'scaffold' (Wood *et al*, 1976 p.90). The process of scaffolding provides, as the following example shows, the initial means by which students are helped to 'see' the phenomena in the same 'scientific way' that the teacher 'sees' it (Ogborn *et al*, 1996):

Dr. Starbeck: [Points to the animated character moving around a stylised circuit on the whiteboard] *'Right, so we've got something moving around a circuit, a person moves around the circuit. What's moving around a real circuit?'*
Student SK4: *'Electrons.'*
Dr. Starbeck: *'Ok, electrons, electric charges. So the person* [points to character on screen] *stands for...?'*
Student SK5: *'Charge.'*
Dr. Starbeck: *'Electrons, charges. People stand for charges* [points to animated

character walking around circuit]. *What do charges do? They move around the circuit. What do we call a movement of charge?'* [this had been taught in a previous lesson].
Student SK14: *'Current.'*
Dr. Starbeck: *'Current. Current, right. So we've got charges moving around a circuit, people carrying boxes, charges carrying energy around a circuit. OK?'*

Indeed, Lunetta (1998) has argued that *'laboratory inquiry alone is not sufficient to enable pupils to construct the complex conceptual understandings of the contemporary scientific community. If pupils' understandings are to be changed towards those of accepted science, then intervention and negotiation with an authority, usually a teacher, is essential'* (p.252).

The issue then is the extent to which this intervention and negotiation with the teacher is acknowledged and built into the practical task by the teacher.

The role of language in practical work

Getting students to think about objects and observables using particular scientific ideas and words can be difficult (see also Chapters 16 and 17). Part of the reason for this is that scientific language can itself involve either unfamiliar and/or strange words (e.g. 'photosynthesis', 'titration' and 'kinetic') or, even if the words are familiar, they can have alternative, well established, non-scientific meanings widely used in everyday life (e.g. 'cell', 'mixture' and 'work'). Indeed, it has been reported (Fotou & Abrahams, 2016) that the use of such scientific language, without proper explanation and guidance by the teacher, can itself cause further confusion amongst some students. An example of an effective intervention and negotiation can be seen in the above example, in which Dr. Starbeck scaffolds the use of scientific terminology. It is the encouraging of students to *use* and *understand* scientific language when talking about their scientific observations and ideas that enables them to become, in a Kuhnian (1970) sense, practising members of the scientific paradigm.

In conclusion

Given the clear importance of trying to ensure that students do what the teacher intends with objects and materials in the limited time available, 'recipe'-style tasks are likely to continue to play a prominent role in science practical work. However, if the scale of the cognitive challenge that students can sometimes face in linking their actions and observations to a framework of ideas were better recognised, teachers might start to divide the time in a practical lesson more equitably between 'doing' and 'learning'. Such a separation would see a greater proportion of the lesson time being devoted to helping students to use ideas, concepts and scientific language associated with the phenomenon that they want them to produce, rather than simply seeing the production of the phenomenon as a successful end in itself.

Further reading

Abrahams, I. (2011) *Practical work in school science: A minds-on approach.* London: Continuum

Abrahams, I. & Reiss, M. (Eds.) (2016) *Enhancing learning with effective practical science 11-16.* London: Bloomsbury

Osborne, J. & Dillon, J. (2010) *Good practice in science teaching: What research has to say.* Maidenhead: Open University Press

References

Abrahams, I. & Millar, R. (2008) 'Does practical work really work? A study of the effectiveness of practical work as a teaching and learning method in school science', *International Journal of Science Education*, **30,** (14), 1945–1969

Arons, A. (1993) 'Guiding insight and inquiry in the introductory physics laboratory', *The Physics Teacher*, **31**, 278–282

Bell, J., Bush, T., Fox, A., Gooddey, J. & Goulding, S. (Eds.) (1984) *Conducting small-scale investigations in educational management.* London: Harper and Row

Driver, R. (1975) 'The name of the game', *School Science Review*, **56,** (197), 800–805

Fotou, N. & Abrahams, I. (2015) 'Doing with ideas: The role of talk in effective practical work in science', *School Science Review*, **97,** (359), 55–60

Kuhn, T.S. (1970) *The structure of scientific revolutions.* Chicago: Chicago University Press

Leach, J. & Scott, P. (1995) 'The demands of learning science concepts: Issues of theory and practice', *School Science Review*, **76,** (277), 47–52

Lunetta, V.N. (1998) 'The school science laboratory: Historical perspectives and contexts for contemporary teaching'. In *International Handbook of Science Education Part 1*, Tobin, K. & Fraser, B. (Eds.). Dordrecht: Kluwer

Millar, R., Le Maréchal, J-F. & Tiberghien, A. (1999) '"Mapping" the domain: Varieties of practical work'. In *Practical work in science education*, Leach, J. & Paulsen, A. (Eds.). Dordrecht: Kluwer

Ogborn, J., Kress, G., Martins, I. & McGillicuddy, K. (1996) *Explaining science in the classroom*. Buckingham: Open University Press

Solomon, J. (1988) 'Learning through experiment', *Studies in Science Education*, **15**, 103–108

Solomon, J. (1999) 'Envisionment in practical work. Helping pupils to imagine concepts while carrying out experiments'. In *Practical work in science education – recent research studies*, Leach, J. & Paulsen, A. (Eds.), pps. 60–74. Roskilde/Dordrecht: Roskilde University Press/Kluwer

Tiberghien, A. (2000) 'Designing teaching situations in the secondary school'. In *Improving Science Education: The contribution of research*, Millar, R., Leach, J. & Osborne, J. (Eds.). Buckingham: Open University Press

Wallace, G. (1996) 'Engaging with learning'. In *School improvement: What can pupils tell us?*, Rudduck, J. (Ed.). London: David Fulton

Wood, D., Bruner, J.S. & Ross, G. (1976) 'The role of tutoring in problem solving', *Journal of Child Psychology and Psychiatry*, **17**, 89–100

Woolnough, B.E. & Allsop, T. (1985) *Practical Work in Science*. Cambridge: Cambridge University Press

CHAPTER 13

Taking science teaching outside the classroom: using a pedagogical framework

Melissa Glackin & Natasha Serret

This chapter briefly considers the current situation for teaching and learning outside the classroom within the UK and begins to challenge some of the existing definitions and beliefs about outdoor learning. These perceptions of outdoor learning can sometimes contribute to the barriers that dissuade teachers from using the outdoors as a learning context. Some research findings from 'Thinking Beyond the Classroom' (PSTT, see Websites) are described. The programme's pedagogical framework for teaching and learning outdoors forms the backbone of this chapter. The chapter also highlights some practical advice to support managing learning outdoors.

The importance of teaching and learning outside the classroom

The school laboratory traditionally has been the chief venue for science learning. However, over the last decade, there has been a gradual rise in profile of teaching science in differing locations. In parallel with a growing body of research that suggests the many benefits of learning outside the classroom, past and present policy directives and government reports have promoted taking students beyond the classroom.

In England, over a decade ago, the Department for Children, Schools and Families (now the Department for Education) launched the manifesto for Learning Outside the Classroom (LotC) (DCSF, 2006). The manifesto stated that all children educated in England should gain direct experience of learning in different contexts. The manifesto's legacy is evident in the establishment of the Council for Learning Outside the Classroom; a body charged to promote outdoor learning whilst raising the quality of experience offered (see lotc.org.uk). More recent government support for outdoor learning has come from the Department for Environment, Food and Rural Affairs (DEFRA) and Natural England who, in 2016, announced that every school student in England will have the opportunity to visit the National Parks at each stage of their education (DEFRA, 2016).

In Scotland, learning outside the classroom is considered central to a rounded education. That is, the newly revised 'Curriculum for Excellence' establishes outdoor learning for students as a mandatory requirement, situated alongside education for sustainable development and global citizenship (Education Scotland, 2017).

Redefining the role of outdoor learning in science education

Throughout this chapter, we hope to challenge perceptions about the nature and purpose of outdoor learning, and the often traditionally held view that the outdoors is for ecology fieldwork alone. To this end, this chapter is exemplified with accounts from an innovation programme with which both authors have both been involved, *Thinking Beyond the Classroom*. The programme aimed to develop the skills and confidence of science teachers when teaching outdoors, alongside the creation of ten lesson activities.

Often, learning outside the classroom is referred to as 'informal learning'. We believe that the use of such a term is misleading, as learning in or outside the classroom can be equally formal and structured in nature. In our view, the outdoors is simply an environment for learning. As with teaching inside the classroom, if these outdoor opportunities are well planned and taught, they can strengthen science learning, providing cognitive development and curriculum-related outcomes, as well as giving depth to the curriculum and contributing to students' physical, personal and social education (Rickinson et al, 2004).

Further, science learning outside the classroom has the potential to offer access to rare materials, the possibility of extended and authentic practical work, an introduction to the 'messiness of science' and the improved development and integration of scientific concepts (Braund & Reiss, 2006). These aspects need not be confined to the teaching and learning of ecology and, throughout this chapter, we illustrate ideas with examples of how the outside space can be used to develop understanding in concepts such as 'properties of matter' and 'forces', topics often aligned with chemistry and physics.

Opportunities for learning science outside the classroom can take place in a variety of contexts and through a range of activities. These include experiences further afield, such as fieldwork abroad, visits to the wider countryside and sites of industry. In addition, they include excursions to farms, zoos, botanic gardens, museums, science centres and cultural sites. However, there is also a wealth of potential in immediate local sites, such as parks, city squares and the school grounds. In terms of duration, students might spend 15 minutes collecting observations in the school grounds, a double period spent in a park, a day visit or a residential experience. In relation to the nature of the programme to be discussed, this chapter primarily concentrates on learning in the context of a space outside that can be accessed quickly from the science laboratory. To this end, regardless of how experienced a science teacher may be, s/he should be able to trial some of the ideas we discuss.

In this chapter, we present an overview of *Thinking Beyond the Classroom*, and draw from the findings to discuss some of the features that appear to enable effective science learning in the local outdoors. In so doing, we consider some of the practical aspects involved in planning and teaching outdoors and also offer some examples of Key Stage 3 (KS3, ages 11-14) lessons that were developed to maximise the opportunities offered by the space outside the classroom.

A pedagogical framework that supports teaching and learning outdoors

Thinking Beyond the Classroom

A joint King's College London and Field Studies Council programme (funded by the AstraZeneca Science Teaching Trust, now Primary Science Teaching Trust) was established to explore ways of tapping into the science learning potential of local urban outdoor contexts. The teachers involved in the professional development programme all taught in secondary schools and were drawn from a range of science backgrounds and experiences of teaching outdoors.

Drawing from current educational research, elements of Cognitive Acceleration and Assessment for Learning were incorporated into a pedagogical framework. This framework provided the structure for the set of ten KS3 activities and the professional development programme that emerged from this project. Below, we show how the core aspects of this pedagogical framework provide the foundations of an approach that supports teaching and learning outdoors. The website address from which to access the complete set of activities and the professional development programme can be found at the end of this chapter.

Observing the local environment

Observation is one of the key processes that scientists use in scientific inquiry. In science, observing requires someone to do more than simply look at something. Crucial to this process is the development of students' appreciation of the difference between looking and observing. It is multi-staged and complex and requires being open to noticing what is around one and progresses to being able to focus attention on a particular feature, using all one's senses to describe it in detail. This process equips us with information to help us to name, explain and document what we see, enabling us to identify what is familiar and what is unusual. Effective observation triggers more questions, which can lead to further investigations (Eberbach & Crowley, 2009).

Assessing and tracking progress in observation is not straightforward. The skill of observation is barely recognised in national tests. This seems strange, considering eminent scientists' findings have relied on excellent skills of observation. For example, Charles Darwin's observations on the Galápagos Islands, which led to his theory of evolution, reinforce an appreciation that quality observations develop over time, require a great deal of patience and are enriched through repetition in a range of contexts.

Working with our participating teachers and their students, we have learnt that enabling students to observe at this level in the outdoors can be challenging. Set within a familiar context such as the school grounds, observation work may not appear as appealing compared with a new, exciting environment where noticing is a natural/instinctive behaviour. Furthermore, students are unused to exploring the everyday through the lens of scientific observation. However, as Frøyland and colleagues (2016) found, students who are taught the skills of observation have enhanced recall and scientific knowledge compared to those who are not taught such skills.

The following considerations and activities provide some structure to help develop students' ability to observe within a local setting:

➢ Start by encouraging students to explore an everyday feature in a variety of ways and consider the science that might underpin what they see: for example, rusting around a gutter.

➢ Challenge students to go beyond a one-word description. Encourage students to imagine that they are talking to someone who can't see or isn't familiar with what they are observing. Focus their attention on the colour, texture, the pattern of its distribution and age of the rust.

➢ Widen the frame/boundaries of this observation and extend their observation to look at the rust in its wider context. Consider where the rust is and where it isn't, the structures on which it occurs and the clues that these wider observations might elicit.

➢ Use questions to promote repeated observation so that students can begin to see something familiar through different science lenses.

➢ In your planning, consider what kinds of observations they will make. What will they need for this?

➢ Encourage the use of the senses before using technical equipment. Think about the selection of resources for observations (e.g. magnifying glasses); do they enrich or distract from the observation experience?

Collaborative group work

We make meaning together, by working and talking together and exchanging ideas (Vygotsky, 1986; Mercer & Dawes, 2014). It is a critical element in the highly successful Cognitive Acceleration through Science Education (CASE) project. One of the key pillars in CASE theory (Shayer & Adey, 2002) is social construction, which is rooted in a Vygotskian (Vygotsky, 1986) theoretical framework that emphasises social interaction in cognitive development and learning. All the lesson activities that we have developed as part of the *Thinking Beyond the Classroom* programme promote collaborative, independent group work by fostering a supportive environment where ideas can be generated and challenged. However, using group work as a tool for learning inside and outside the classroom requires some foundation work: students need to have opportunities to establish some particular ground rules prior to outdoor group work, to develop an awareness of the role of negotiation, delegation and decision-making and to appreciate that group work improves with reflection and time (Barnes, Rubie-Davies & Blatchford, 2009).

Organising group work outside can be a daunting prospect. Group work tends to be more unstructured, chaotic and unpredictable compared with individual or whole class teaching. Large class sizes, limited equipment, behaviour and attainment can influence a teacher's decision over whether to use group interaction in learning activities. Research (for example, Webb, 2009; Kutnick & Blatchford, 2014) argues that teachers need more professional support in how to establish group work and enable students to manage their groups independently. For our participating teachers, taking group work from the classroom into the outdoors was an additional challenge in itself (Glackin, 2017). However, we observed that effective group work is a crucial aspect of learning science outdoors and, with time and support, can enrich an outdoor science experience. We found that effective groups:

➢ are small in size, ideally threes or fours, are mixed ability and have a mix of males and females;

➢ contain a number of personalities that all feed into the group's potential success (see Box 1);

➢ are aware of the different roles needed to enable group work;

➤ have agreed some ground rules and can refer back to these if they encounter a problem;

➤ are supported by their teacher and others in the class in how to negotiate and delegate;

➤ have talked about how to reach a decision when a number of conflicting ideas are suggested;

➤ are constantly encouraged to reflect on how well they have worked as a group, identify what went well and what they can improve on next time; and

➤ have received constructive feedback from their peers and their teachers that focuses on group work skills as well as their understanding in science.

Box 1: Personalities contributing to successful group work

A number of personalities feed into the group's potential success. For example, it is useful to have someone in a group who is a 'risk-taker/initiator' and can enthuse the group to pursue an idea that they might not be sure about at the start. Equally important is the need to have an 'encourager', someone who makes sure that everyone is involved, and a 'peacemaker', someone who can diffuse tensions where the group's collective ideas clash. Another crucial role is that of the 'clarifier/checker'. Although the group may feel that this member is holding the group back because s/he typically asks a lot of questions to help clarify the task, a checker plays a crucial part in the group's success. S/he often asks the questions that other members are afraid to ask and, in doing so, creates opportunities for the group to strengthen their shared understanding.

Provoking cognitive challenge

How can we encourage high-level thinking outdoors? Drawing from studies of Cognitive Acceleration (Shayer & Adey, 2002), teachers with whom we have worked have identified the outdoor context as an excellent vehicle for the promotion of cognitive challenge – the experience of a phenomenon or an alternative idea that is often unexpected and leads to the construction of further understanding (Glackin, 2016). During the programme, for example, we found that a number of alternative ideas emerged when we showed students a range of images from their local environment (such as lichen on the pavement, pigeons in the playground, a puddle of water on the pavement) and asked them to decide which pictures showed habitats. These conflicting ideas are then explored in the context of an investigation in the local environment by asking '*How do you decide if somewhere is a habitat? Do you think there are any habitats in our school? What would you look for, what kinds of data would you collect, to be certain that it is a habitat?*'.

Provoking cognitive challenge in this way is only the start of the learning process. Students need to have the opportunity to take what they have observed outside, the ideas that have been shared and any experiences of conflict, and try to apply this to a related question. According to research into children's thinking, cognitive development is secured through metacognition, an opportunity for the learner to consider how his/her thinking has been challenged and transformed (Shayer & Adey, 2002). Within our programme, we allocated time specifically for this, wherein the students reflected on how they have arrived at this reconstructed level of science understanding. During this time, students also considered how interactions within their group enabled or hindered this process. This process gives them an insight into how their science understanding evolved through interacting with their local environment and with each other.

Learning through questioning

Having organised the groups, set the scene and provided cognitive challenge, the teacher has another role. A teacher can use questions to open up the opportunity for students to reveal what they know and challenge them to explain their thinking (as outlined in *Science inside the Black Box* (Black & Harrison, 2004)). Student questions can be equally essential and ultimately the autonomous student can steer his/her learning through his/her questions. The outdoor context can be a fantastic stimulus, giving students the motivation and confidence to ask their own questions of what they observe around them. However, sending students off outside with the task of 'coming up with a question that can be investigated using first-hand data collection' is a challenge, and can feel risky, for even the most seasoned professional scientist (Glackin, 2016). The process of turning initial thoughts and responses into questions and then transforming these into questions that can be tested scientifically is complex and challenging for adults as well as students. Students might need time in the classroom looking at questions related to science in the outdoors and identifying which ones can be investigated. Harrison and Howard (2010) offer useful question learning stems, which provide support for both teachers and students to create their own questions such as 'Is it always true when…?' and 'How are they the same and how are they different?'

Some examples of questioning in the outdoors that our teachers found useful were:

➢ If you left a cup of water outside for a week, what could happen? Why?

➢ How many different examples of forces might you spot around you on your journey between home and school? How can you tell if these forces are balanced or not?

> ➤ Did you have any unexpected findings?
> ➤ Did you change anything about the way you observed your site on the second visit? What did you do differently? Why?

Managing teaching and learning outside

Teaching students outside might initially seem a daunting prospect. Teachers often report that their initial hesitance stems from their own lack of familiarity with the school grounds, their confidence to manage students' learning outside the confines of the classroom, and the potential health and safety implications (Lock, 2010). It is worth stressing that the teachers with whom we have worked all initially had some reservations about teaching outside and were often more controlling of student learning when in the new context compared to the classroom. However, gradually, as the teachers brought their students outside more often, they began to feel more confident to experiment with the core pedagogical strategies in their practice. Hence, teaching effectively outside takes time and patience (Glackin, 2017).

In the section below, we offer practical advice resulting from lesson observations and teacher discussions. The advice focuses on aspects for consideration prior to the lesson, and potential teaching and management strategies during the lesson:

Before the lesson
Consider some of the less obvious areas within the school grounds, as these offer excellent sources of scientific phenomena. These alternative sites, in addition to the school playground or school pond, may be less familiar to pupils, and may be less overlooked by other classrooms and therefore hold fewer distractions.

Check with the Head of Department and the Education Visits Co-ordinator (EVC) about possible relevant school policies and documentation. The distance of the site from the classroom and the route that will be used to access it should be planned. Offering students a plan/base map of the school grounds enables them to orientate themselves and can be used as a resource to record site observations. A good source for site maps is Google Earth. Deciding where invisible line boundaries will be drawn, and an appropriate fixed space for gathering and teaching, will encourage the class to see the space as an 'outdoor laboratory' where school rules still apply. Such fixtures could also be illustrated on the map.

Prepare a 'Grab-and-go sack'. This includes all the resources needed for working with students outdoors, such as:

➢ an alternative outdoor whiteboard (for example, laminated A3 paper/or poster paper), which can be mounted at the designated gathering space (such as on a brick wall, large tree);

➢ thick felt-tip pens;

➢ mini-first aid kit;

➢ antiseptic gel (an excellent alternative to the whole class needing to wash their hands when returning to the classroom);

➢ tissues and wet-wipes;

➢ a set of pencils and A4 paper;

➢ camera(s) on phones or iPads (photographs taken with iPads can be drawn on directly using software such as *freezepaintapp.com*);

➢ plastic sitters (carrier bags are an ideal alternative);

➢ a whistle;

➢ laminated 'learning through question-stems' (see above, Harrison & Howard, 2010);

➢ rubbish bags; and

➢ plastic disposable gloves (for pupils who need encouragement to touch outdoor objects).

Considerations during the lesson

Weather can be an obstacle but, with a little bit of planning, can be overcome. Teachers with whom we have worked ask students to bring in rain jackets and appropriate footwear for the next lesson. Building up a supply of spare wet weather gear may also be useful (from lost property or second-hand shops).

Remind pupils who have medical conditions, such as asthma, to carry their medicines. Students within their groups need to negotiate outdoor ground rules. If inappropriate behaviour occurs, asking students to reflect on their rules for productive working will reinforce the goal of social learning. Separating equipment into group sets enables quick distribution. Identifying a timekeeper within each group, equipped with a stopwatch and knowledge of the designated gathering space, enables groups to be gathered together with ease. At the end, gather the class, get groups to check equipment and collect any litter. If lesson plenaries are conducted outside, students could be dismissed from the site, instead of returning to the classroom. This will enable more learning time and encourage the students to refer to outside examples in the plenary discussion. Teachers report that, as students work more frequently outside the classroom, they begin to establish the required learning behaviours.

In conclusion

Science happens outside the classroom and local outdoor contexts can help to make science more relevant and accessible for students. Our research suggests that the processes involved when deciding what and how to teach outside are informed by the same pedagogical expertise required for effective teaching inside the classroom. This chapter has set out how research-informed science education teaching strategies used inside the classroom can be effectively used outside the classroom. That is, we have used aspects from theories of Cognitive Acceleration and Assessment for Learning, such as questioning, group work and provoking cognitive challenge, to develop activities that can be taken outside. This chapter particularly focused on how to use immediate and local outdoor contexts as the starting point for developing science understanding across all science subject areas, not just biology. We hope that this approach helps to challenge the traditionally held view that learning outdoors is merely and solely suitable for ecology data collection. The suggestions in this chapter encourage an integration of short, regular outdoor learning opportunities within everyday practice, so that outdoor learning is not simply associated with extended, one-off residential field visits. In this way, learning outdoors can help students to recognise the links between everyday phenomena and the science concepts presented in formal classrooms or textbooks.

Further reading

Amos, R. & Reiss, M. (2012) 'The benefits of residential fieldwork for school science: Insights from a five-year initiative for inner-city students in the UK', *International Journal of Science Education*, **34,** (4), 485–511

Thinking Beyond the Classroom: https://pstt.org.uk/resources/cpd-units/thinking-beyond-the-classroom

Websites

Thinking Beyond the Classroom:
    https://pstt.org.uk/resources/cpd-units/thinking-beyond-the-classroom

Freeze Paint: http://www.freezepaintapp.com/

Council for Learning Outside the Classroom: http://www.lotc.org.uk/

References:

Baines, E., Rubie-Davies, C. & Blatchford, P. (2009) 'Improving pupil group work interaction and dialogue in primary classrooms: results from a year-long intervention study', *Cambridge Journal of Education*, **39,** (1), 95–117

Black, P. & Harrison, C. (2004) *Science inside the Black Box*. London: nferNelson

Braund, M. & Reiss, M. (2006) 'Towards a more authentic science curriculum: the contribution of out-of-school learning', *International Journal of Science Education*, **28,** (12), 1373–1388

Department for Children, Families and Schools (2006) *Learning Outside the Classroom Manifesto*, London: DCFS

Department for Environment, Food and Rural Affairs (2016) *8-Point Plan for England's National Parks*. DEFRA publications. Available at: www.gov.uk/government/publications Accessed 12.04.17

Eberbach, C. & Crowley, K. (2009) 'From everyday to scientific observation: How children learn to observe the biologist's world', *Review of Educational Research*, **79,** (1), 39–68

Education Scotland (2017) *Home page*. Available at: https://education.gov.scot/ Accessed 12.04.17

Frøyland, M., Remmen, K.B. & Sørvik, G.O. (2016) 'Name-dropping or understanding?: Teaching to observe geologically', *Science Education*, **100,** (5), 923–951

Glackin, M. (2016) '"Risky fun" or "Authentic science"? How teachers' beliefs influence their practice during a professional development programme on outdoor learning', *International Journal of Science Education*, **38,** (3), 409–433

Glackin, M. (2017) '"Control must be maintained": exploring teachers' pedagogical practice outside the classroom', *British Journal of Sociology of Education*, 1–16

Harrison, C. & Howard, S. (2010) *Inside the Primary Black Box: assessment for learning in primary and early years classrooms*. London: GL Assessment

Kutnick, P. & Blatchford, P. (2014) 'Groups and classrooms'. In *Effective group work in primary school classrooms*, (pps. 23–49). Netherlands: Springer

Lock, R. (2010) 'Biology fieldwork in schools and colleges in the UK: an analysis of empirical research from 1963 to 2009', *Journal of Biological Education*, **44,** (2), 58–64

Mercer, N. & Dawes, L. (2014) 'The study of talk between teachers and students, from the 1970s until the 2010s', *Oxford Review of Education*, **40,** (4), 430–445

Rickinson, M., Dillon, J., Teamey, K., Morris, M., Choi, M.Y., Sanders, D. & Benefield, P. (2004) *A review of research on outdoor learning*, Shrewsbury: Field Studies Council

Shayer, M. & Adey, P. (2002) *Learning Intelligence: Cognitive Acceleration across the Curriculum from 5 to 15 years*. Buckingham: Open University Press

Webb, N.M. (2009) 'The teacher's role in promoting collaborative dialogue in the classroom', *British Journal of Educational Psychology*, **79,** (1), 1–28

Vygotsky, L.S. (1986) *Thought and Language*. (Kozlin, A., Ed. and trans.) Cambridge, MA: MIT Press

CHAPTER 14

Mathematics in science teaching

Richard Needham & Mike Sands

This chapter looks at how and why mathematics is central to good science teaching and recognises the difficulty that science teachers may have in teaching mathematical skills to their students. It highlights some common problem areas for students and, from the findings of a recent case study, suggests ways of tackling these issues in the classroom.

The importance of mathematics in learning science

Mathematics is an essential component of doing and learning about science. Specifically:

➢ Mathematics provides a language to describe phenomena and relationships between phenomena;

➢ Mathematics enables measurements to be made, recorded and analysed, to detect patterns, to provide explanations and to make predictions; and

➢ Mathematics supports the development of scientific models, which in turn can lead to an improved understanding of the world around us.

Many secondary students regard science and mathematics as two different curriculum subjects, without appreciating the vital role of mathematical understanding in the science curriculum.

Mathematics in the 11-16 science curriculum

From 2016, all English science GCSEs contain a common list of mathematical skills to be included in the science curriculum, along with defined contexts in which each skill is to be used. Welsh and Northern Irish science specifications contain very similar lists. So, although many of these skills were requirements of previous specifications, they are now more tightly defined and have been given greater prominence. In addition, there is now an obligation to assess each of these skills, and to award a specified minimum number of marks (which varies between different disciplines) in science examinations to questions containing mathematics.

In Scotland, the picture is slightly different, but there is still a focus on mathematics. Every teacher has a responsibility for the teaching of numeracy and mathematics to students up to age 15.

Mathematical skills and support from ASE

As part of their science curriculum, 16 year-olds are to be taught 25 different mathematical skills. Current curriculum documents, such as the Department for Education's *GCSE subject content for combined science*, group these skills under five mathematical headings: *Arithmetic and computational skills, Handling data, Algebra, Graphs, Geometry and Trigonometry*. Unfortunately, some of these headings, and the skills within them, use language not familiar to science teachers and so need some interpretation and explanation.

ASE set up its *Language of Mathematics in Science* project in 2014 in response to concerns from ASE members, which included the difficulties that some students experience when applying their mathematics learning to science lessons. The project, which was funded by the Nuffield Foundation, aims to support teachers by developing a common understanding of important terms and techniques, and sharing experiences of applying these in science lessons. The project has so far resulted in numerous professional development activities, and the production of two documents:

> ➢ *The Language of Mathematics in Science: A guide for teachers of science 11-16*. This guide provides an overview of relevant ideas in secondary school mathematics, and where they are used in science. Key mathematical terms are explained along with a glossary giving definitions. It indicates where there may be problems in understanding, and identifies some of the approaches used in mathematics teaching that could influence what is done in science.

> ➢ *The Language of Mathematics in Science: Teaching approaches*. This contains accounts by different teachers, describing activities that they have undertaken in response to mathematical problems experienced by their students in science lessons. Some of these could be categorised as individual interventions, others are examples of collaborations between teachers, or whole school initiatives. The booklet outlines the problems faced, the solutions that were used and the range of outcomes achieved.

Both project documents are free to download; details are provided at the end of this chapter.

Using mathematical ideas in science

The project view is that mathematical skills should not be taught in isolation, but taught by science teachers in a scientific context. For this reason, the *Guide for teachers of science 11-16* groups mathematical ideas and terms into areas of scientific activity, as follows:

- ➤ The nature of data and how it is collected;
- ➤ Different ways of representing quantities, and how quantities can be calculated;
- ➤ Representing data for different purposes;
- ➤ How to construct good charts and graphs;
- ➤ Exploring relationships in data, including rates of change and correlation;
- ➤ Understanding proportionality and how it links to ratio, percentage and scale;
- ➤ The causes of variability in data, and how to deal with uncertainty;
- ➤ Using scientific models and mathematical equations; and
- ➤ How mathematics can be used to generate useful scientific information.

These areas do not imply a specific teaching sequence, and each area is likely to be visited many times as students progress through their secondary science education. In some schools, it has been possible to organise the teaching of mathematics and science so that science teachers can build on a particular skill after it has been taught within a mathematics lesson. However, this is not possible in most schools, and it may not be ideal for some of the reasons described in the next section

Mathematical skills and the teaching of mathematics in schools

The *Teaching Approaches* booklet documents numerous examples of students being unable to apply mathematics in science lessons. To reduce some of the learning difficulties for students, it is recommended that science departments work towards building consistency in their approach to mathematical skills within their teaching, and aim to develop consistency in language and approach between science teachers and mathematics teachers in their schools.

The next section looks at consistency between science teachers in more detail. But, in aiming for consistent approaches, it is worth pointing out that differences inevitably exist between the way mathematicians use mathematics and the way that mathematics is used in science. This is one of several reasons why we recognise that science teachers cannot expect their mathematics colleagues to provide students with all the skills needed in science. Others include:

> Curriculum structure – depending on setting arrangements, some students may not be taught the mathematical skills required for higher tier science GCSEs;

> Mathematical language – some vocabulary use in science has a much more specific meaning than when the same word is used in everyday speech. Examples include 'force', 'energy', 'particle' and 'food'. The same is true in mathematics, where some words have a more precise meaning than when the same word may be used in science. Examples of words that may have a different meaning in science and mathematics include 'line', 'histogram', 'range' and 'variable';

> The use of variables – in mathematics, students will be familiar with symbols such as x and y or a, b and c to represent variables in equations. These variables represent a value (number). In science, the symbols used are often specific to the context, such as V, I and R in electricity, or F, m and a in mechanics, and they each convey a precise meaning. These variables often represent a quantity and therefore consist of a value (number) and a unit; and

> The use of graphs – graphs in science lessons are often used to explore the relationships in measured or observed data, involving high levels of variability and complex numbers. This adds to the challenge of graph drawing for science students, compared to the graphs they first encounter in their mathematics lessons.

The need for consistency

In general, the project found that science teachers are confident users of the mathematical skills required when teaching secondary science. However, there is little consistency in the way that different teachers use these skills. As an illustration, when science teachers were asked in a workshop to draw a graph from some data, the outcomes were very varied. The imaginary data provided showed how much product was produced by a fermentation reaction at different time intervals. Teachers were provided with a sheet of graph paper, a pencil and a ruler, and asked to draw the graph that they would expect from a good GCSE candidate, adding a line of best fit.

Example 1 of a teacher-drawn graph

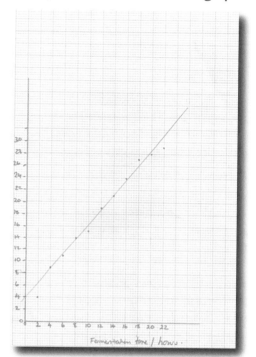

Example 2 of a teacher-drawn graph

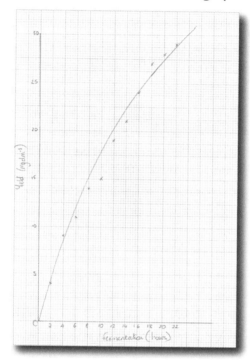

Comparison of the two teacher-drawn graphs		
Feature	Example 1	Example 2
Vertical scale	1 cm = 2 units	1 cm = 1.25 units
Vertical axis units	mg (dm³)	(mg dm⁻³)
Horizontal axis units	time / hours	(hours)
Co-ordinate marker	dot	diagonal (saltire) cross
Line of best fit	straight	curve
Intersection with axes	vertical axis = 4	origin

The illustrations shown are just two examples of the variability found in this exercise; there is very little consistency in outcomes between science teachers when they are asked to draw graphs in similar workshops. As a result, we suggest that all science teachers within a school should agree some common approaches to graph-drawing and other mathematical skills, to reduce opportunities for confusion amongst their students.

Our recommendations for graph drawing are:

➤ Scale – restrict the number of scaling options, and give inexperienced students access to pre-drawn templates illustrating the scales available. For example, the only permissible scales could be 1 cm = 1 unit, 1 cm = 2 units or 1 cm = 5 units (or multiples of 10);

➤ Units – units of distance in metres could appear as: *distance in m*; *distance (m)*; *distance / m*. Our preference is for *distance (m)* up to age 16, then use *distance / m* for older students. This is to avoid confusion when stating units for more complex quantities such as velocity, where writing *velocity (m/s)* is less confusing than *velocity /m/s*;

➤ Co-ordinates – marked by a cross rather than a dot or a dot in a circle;

➤ Line of best fit – whether the line of best fit is a line or a curve depends on the relationship between the specific variables measured in the investigation. Students need to understand the scientific relationship between the variables when making their choice; and

➤ Origin – if there is a proportional relationship between the two variables, then the line should pass through the origin. Students need to understand the relationship between variables before deciding if the origin of the graph can be considered an additional data point.

Graphs serve many different purposes in science, representing a variety of relationships and data types. Consequently, there is no single 'correct' way to draw graphs in science, and so it is important for teachers within a school to discuss, agree and communicate the common approaches to be used. Other mathematical skills where common approaches are helpful include:

➤ Rearranging algebraic equations;

➤ Using unit prefixes or standard form;

➤ The use of scale factors in drawings and diagrams; and

➤ Describing data types.

For further guidance on these and other issues, see *The Language of Mathematics in Science: A guide for teachers of science 11-16*.

Teaching mathematical skills as part of the science curriculum

Three recommendations have emerged through discussions with teachers at *Language of Mathematics in Science* workshops:

> ➤ Embed the teaching of mathematics skills in science lessons throughout the secondary curriculum, building on the foundations created in primary school;

> ➤ Break down complex skills into a series of steps, teaching the steps individually before helping students to apply that skill in a science context; and

> ➤ Use simple science contexts initially to introduce new mathematical skills, then plan for progression in these skills through a variety of science contexts.

In the next section, *Looking at graph drawing*, we describe working with students in a school, focusing on two of the above recommendations: breaking down complex skills into smaller steps, and embedding the teaching of mathematical skills throughout the secondary curriculum.

Looking at graph drawing

In this school, students have major difficulties when drawing graphs in science. We worked with four students (age 11) with good levels of attainment reported by their primary schools, to try to understand the problems that similar students had encountered in the past. We aimed to use this information, when reviewing our schemes of learning, to develop new ways of approaching mathematical skills.

We focused on the following skills:

> ➤ The arrangement of data in tables and identification of variables and scales.

> ➤ The construction of suitable graphs.

> ➤ The interpretation of the graph.

In the initial activity, students were presented with some data using the relationship of distance and time to calculate speed. The data given were simple and increased in even amounts – distance data in increments of 10 and time data in increments of two. There were six data pairs given. The students were asked to put the data into a table, plot a suitable graph and then asked to interpret the graph.

When questioned about which variable would go on the *x*-axis and which on the *y*-axis, responses were usually based upon the way in which the table was drawn, or presented by the teacher. Students had heard the terms 'independent' and 'dependent' variables in lessons, but felt confused by them and knew that the teacher would tell them what to plot on the axes.

From this activity, it was clear that students had issues with the variables and the choice of the axis on which to place each variable.

The next session was planned to look at introducing the terms 'independent' and 'dependent' variable and encouraging students to consistently use comparative '-er' words when describing correlation in the data. The students were given some teaching on the use of comparative phraseology ('-er' words) and told that one variable depended (dependent variable) upon the other one (independent variable). The students were given a mini-whiteboard and asked to identify both the independent and dependent variable as well as use '-er' words to describe the relationship: for example, 'the long**er** the time, the great**er** the distance'.

Now that the students had an idea of how to identify variables and describe simple relationships, the final session focused on more complex data and relationships. Data from a distance-time graph were used, which included a stationary period so that the graph was not a continuous straight line. The intervals between data points did not follow a regular pattern, so the correct scale needed to be chosen.

Initially the students were unable to draw the graph. Although they could now recognise the variables and correctly identify which variable went on each axis, the issue was choosing the scales. When offered the opportunity to use a scale selector tool (see Goldsworthy *et al*, 1999), they found scaling much easier to carry out.

This short three-session study identified several areas where we needed to adjust our schemes of learning:

1. When teaching graphing skills, students need to be able to identify variables and to describe their relationship from the start and these skills need to be revisited at every opportunity.
2. Students need to become confident at drawing graphs using explicitly taught graph-drawing skills. These should be taught separately from ways of analysing relationships, to avoid confusion.
3. Providing students with pre-drawn axes helps them to focus on relationships between variables and their interpretation.
4. Cognitive Acceleration through Science Education (CASE)-style lessons will help to reinforce graph drawing and interpretation skills. We plan to introduce these into our scheme of learning (see www.kcl.ac.uk/sspp/departments/education/research/Research-Centres/crestem/Research/Past-Projects/Cognaccel.aspx).

Summary of learning difficulties and their solution (graph-drawing activity)

Student learning difficulty	Suggested teaching approach
Choosing which data to display on each axis	Teach students to identify dependent and independent variables, leading to the correct design of data tables and then using this table to indicate which data set is displayed on which graph axis.
Describing possible correlations in the data	Ensure that students understand that the dependent variable 'depends on' the independent variable, then use '-er' words to describe the relationship between the data values for each of the variables.
Choice of a scale for each axis	Allow only a limited choice of scale (such as 1 cm = 1, 2 or 5 units only). Present students with a series of pre-drawn scales on similar graph paper to their own from which to choose.

Looking at calculations and equations

In this section, we describe working with a different group of students to explore the third workshop recommendation – how to use a simple context to introduce a new mathematical skill, then build progression through a variety of science contexts.

At a recent staff meeting, we identified calculations and equations as skills that were lacking in GCSE students. In this second study, the authors worked with a group of four 12 year-old students, over three lunchtime sessions, to explore the difficulties that they faced and to investigate teaching approaches.

The first session was designed to find out what the students already knew. They were given similar data to those used in the previous graph drawing study and were asked to plot a graph, draw a line of best fit and then to find the gradient of the line. They were also given the speed equation in algebraic form ($s = d/t$), asked to identify the variables and then make d the subject of the equation. Students could draw the graph, label the axes with units and, in most cases, add a line of best fit. They struggled to calculate the gradient, even though they could tell us the speed of the object. Surprisingly however, they could not relate the equation to the graph that they had just drawn and did not know how to rearrange the equation.

On reflection, it was clear that these students found constructing graphs and drawing the line of best fit just as difficult as the younger students. The older students seemed to have a basic understanding of the relationship between the variables, as some of them could calculate the speed. This suggests that they have a grasp of what the graph showed. The problem calculating the gradient of the line fits with the information from our mathematics department, who explained that this topic was not taught by them until later. Nevertheless, it was of concern that the students were unable to relate the equation to the information given in the data. Perhaps the confusion was caused by the data not being in x and y format. It was decided to consider gradients using pre-drawn graphs and also to rearrange equations using x and y.

In the second session, the students were given a graph similar to the one from the initial exercise but with the units removed. Teaching included explaining that the gradient was the slope of the line and, if the line was straight, we could use this alongside the equation to find something out. Teaching modelled how to draw a triangle from the line to the x axis and then use the simple equation of *change in y / change in x*. They were then asked to work out the answer. Following this, they were given a second graph and asked to calculate the gradient of the line. They all calculated the gradient correctly. We used mini-whiteboards in this activity, as students found this more engaging.

The next step was to write out the *speed = distance / time* equation in word form and the students were asked to compare it to the gradient calculation. This step-by-step approach supported the students in both carrying out and understanding the processes involved. By breaking down the information into manageable 'chunks', the students had seen the connection between the graph, the gradient of the line and the equation for speed.

For the final session, a graph of voltage and resistance was provided, which related to Ohm's law. This time, the word equation was written out with the symbol equations next to it and a reminder of how to calculate the gradient so that the relationships could be seen. The students could successfully calculate the gradient and use it to give a value for the current (we had rearranged the equation for them).

The students were then taught how to rearrange equations using only x and y. For example, $y = 2x$ and $y = 4x$. Values were given for y and the students were asked to calculate x. All students could do this without problems and found it reasonably easy to explain.

We then introduced the speed equation once again and gave the values for speed and time, asking the students to find a value for distance. Once again, they found this more difficult but, eventually, could see the relationship once the similarities between the scientific formulae and the equations involving x and y had been pointed out.

In conclusion, this activity demonstrated several points that we need to consider when reviewing our schemes of learning:

1. Students are more able to use their mathematical skills if a problem is broken down into manageable steps.

2. Language can become a stumbling block when discussing mathematical operations. The students had been confused when we asked them to 'make distance the subject of the speed equation'.

3. By the age of 12-13 years, the students are familiar with x and y so, by using these as variable names rather than 'scientific variables', students are more confident in rearranging equations.

4. When introducing equations into science teaching, give the equation in word form, symbol form and x and y form. Slowly, as confidence increases, remove x and y.

Summary of learning difficulties and their solutions (calculations and equations activity)	
Student learning difficulty	**Suggested teaching approach**
Calculating the gradient of a graph	After being shown how to draw a triangle on the line of best fit, and how to find the change in values of x and y, students calculate change in y / change in x. After this has been mastered, the correct units are introduced
Linking an equation to a graph	Express the equation in word form, not symbols, and explicitly link each term to each axis label.
Rearranging an equation	Introduce scientific words gradually. Express equations in word rather than symbol format, and link the terms in the equation to x and y.

5. The use of specific letters or symbols in some science equations can be confusing. For example, I is used as the scientific variable for current, so knowing that I is from the term *Impedance* can help.

6. Increase difficulty slowly over time, as these are skills that need to be practised regularly.

7. Working with mathematics teachers may help to develop similar use of language and procedures across subjects to reduce confusion and development of misconceptions.

In conclusion

Mathematics is central to both learning and doing science. The recent curriculum changes that reinforce the position of mathematics in science teaching are welcomed, if teachers can be effectively supported to implement these changes.

The *Language of Mathematics in Science* project has made a start in helping to achieve a common understanding of important terms and techniques, and raising awareness of the challenges faced by students. The work done in schools is helping science teachers to adopt common approaches and to talk about different mathematical skills in ways that are consistent with their colleagues. As we have shown in the accounts of this work with students, we now need to develop teaching activities that will enable students to become confident users of mathematics in science. In our next phase, we aim to work with teachers in identifying appropriate contexts for the teaching of mathematics within science and to plan for progression in students' mathematical skills.

Further reading, sources of advice and resources

The Language of Mathematics in Science: A Guide for Teachers of 11-16 Science
The Language of Mathematics in Science: Teaching Approaches
  (Both available for free download at www.ase.org.uk) Accessed 08.17

STEM Learning: A collection of teaching resources:
  www.stem.org.uk/resources/community/collection/23430/mathematics-gcse-science Accessed 08.17

Goldsworthy, A., Watson, R. & Wood-Robinson, V. (1999) *AKSIS: Getting to Grips with Graphs*. Hatfield: Association for Science Education

GCSE subject content for combined science, listing the mathematical skills required for English GCSE science specifications, and the context in which they should be taught. Available at: www.gov.uk/government/uploads/system/uploads/attachment_data/file/593774/Combined_science_GCSE_formatted.pdf Accessed 08/17

Numeracy and mathematics: experiences and outcomes. Scottish Curriculum for Excellence, Education Scotland: www.education.gov.scot/Documents/numeracy-maths-eo.pdf Accessed 09.17

Mathematical skills in triple science – STEM Learning links to guidance on specific mathematical skills: www.stem.org.uk/triplescience/maths Accessed 08.17

Also, guidance and support for the teaching of mathematical skills in science is available from most English awarding bodies. For example:

Edexcel: *Guide to Maths for Scientists* http://qualifications.pearson.com/content/dam/pdf/GCSE/Science/2016/teaching-and-learning-materials/Guide-to-Maths-for-Scientists.pdf Accessed 08.17

OCR: *Mathematical Skills Handbook* www.ocr.org.uk/Images/310651-mathematical-skills-handbook.pdf Accessed 09.17

AQA: *Maths skills in GCSE sciences* www.aqa.org.uk/resources/science/gcse/teach/maths-skills-in-GCSE-sciences

Social media

Many teachers and organisations write about their professional activities, opinions and experiences using social media. An example of a blog being used to share good practice in using mathematics in science is by Neil Atkin, available at: www.neilatkin.com/2017/03/26/ten-ideas-teaching-better-maths-science/

Twitter is also a useful tool for professional development, and discussion of science topics, including the use of mathematics, can often be found at #ASEChat, especially on Monday evenings at 20:00 hours.

Acknowledgements

Thanks to the students of Archbishop Sentamu Academy, Kingston-upon-Hull who took part in our study, for their efforts and for giving up their free time.

CHAPTER 15

Enhancing the teaching of science with ICT

Neil Dixon & Nick Dixon

This chapter considers the importance of using ICT in science learning and teaching. It explores different ways in which ICT might be introduced into lessons to enhance the experiences in the science classroom, giving examples of websites and software that teachers might want to try out. The chapter also discusses the potential benefits of using ICT for both teachers and students.

Introduction

We believe that many science teachers intuitively understand how important Information and Communication Technology (ICT) can be in science teaching to support students' learning. It can also engage and inspire students by providing video clips or animations to help explain difficult topics and bring relevance to obscure areas. ICT can recreate events that could not happen in our science laboratories, either because of time factors, safety or context.

Any reader of a recent GCSE science specification will know that there is a large body of information to be understood, often deeply conceptual and hard to visualise. Effective mastery of knowledge, understanding and application relies deeply on effective communication. It is essential that ICT is not seen as a direct and cheaper replacement for hands-on science practicals, but as an enhancement of them (and many other aspects of our teaching).

Over recent years, most teachers have included more ICT in their science teaching in a variety of ways. This is testament to the high value that many science teachers place on ICT. This said, the Education Endowment Foundation (EEF) reviewed evidence on the impact of digital technologies and concluded that *'Overall, studies consistently find that digital technology is associated with moderate learning gains… However, there is considerable variation in impact. Evidence suggests that technology should be used to supplement other teaching, rather than replace more traditional approaches'* (EEF, 2017 p.1). The question, therefore, is to consider which aspects of ICT support science learning.

Of course, showing clips to our students is not the only way in which we can use ICT in our teaching. Webb (2005) suggests that ICT can support learning in four ways:

➤ Promoting cognitive acceleration;

➤ Enabling a wider range of experiences, so that students can relate science to their own and other real world experiences;

➤ Increasing students' self-management; and

➤ Facilitating data collection and presentation.

For students, ICT can support their literacy and language acquisition in science. Because our students are required to learn *information*, then *communicate* this in a variety of ways (most obviously in their exams) and are, on many occasions, better than or as good as teachers in using *technology*, it does seem sensible to us then that ICT should be both engaging and important in their learning. Students can use ICT for research, to test their understanding with interactive quizzes, and to collect and analyse data from experiments with data logging equipment and software. We hope to confirm these claims in this chapter, give examples of practice and resources that we have used in our own teaching, and describe the potential learning. It is useful to make a point of sharing how to find any clip with students, so that they can watch it at home, encouraging them to share the clip with their parents to stimulate discussion about schoolwork. The URL can be e-mailed to students or, more easily, the search terms can be shared. Students can also be asked to compile a list of these online resources for revision.

Teaching science using ICT: a teacher's perspective (why and how)

In the 1970s and 80s, TVs and video cassette recorders (VCR) were wheeled into science lessons on a large trolley at the end of a topic, so that the class could enjoy a lengthy revision film. However, we are not convinced that this significantly deepened students' learning of science. Nowadays, shorter clips are more often shown at critical points during a lesson, via a digital projector on a large screen (sometimes an interactive whiteboard), and teachers often use the slider at the bottom of many clip-playing programmes to fast forward to and select appropriate points. Interspersing sequences of learning activities in this way allows teachers to focus on particular aspects of the learning for consolidation and discussion.

Short clips are now easily available online, and *YouTube* (www.youtube.com) is an obvious first choice for many teachers. However, much time can be spent filtering through many poor quality amateur clips to find something relevant, but there are gems (e.g. *That's Metal* by Science with Tom). Of course, scientifically accurate and high quality clips can be found by searching on channels belonging to organisations such as the BBC (for example, search 'BBC Teach'), Royal Society of Chemistry, Royal Society of Biology, Royal Institution and Royal Society, as well as many charities, including Science and Plants in Schools (SAPS).

It is essential that teachers carefully consider what the use of ICT will add to the learning experience when planning, and consider if it offers new learning opportunities. After identifying the objective, the next step is to match the correct technology (McFarlane & Sakellariou, 2002). Other considerations for using ICT in lessons include whether it is age-appropriate and whether it provides a balanced viewpoint if dealing with an ethical issue. Through an ICT approach, students can access a vast amount of data and research and be introduced to a range of viewpoints. The problem is helping students sift and select those articles that are relevant and that interpret the data in a reliable and scientific way. The Biochemical Society has developed an excellent resource for dealing with some ethical issues in biology at http://www.sciberbrain.org. This offers resources for both GCSE and A-level and helps teachers to support students in making informed decisions about controversial issues.

Experiences that cannot be created in the lab

As previously mentioned, science curricula contain a number of topics that are very difficult to demonstrate in the lab. The website, https://www.brainpop.co.uk, possesses a huge number of short animations with interactive quizzes for students to take. Tim and his robot companion Moby are very popular with many science classes. This website has many free resources, but requires a login and subscription to access their full and extremely extensive library of animations. https://phet.colorado.edu/ is an alternative free site, but without quizzes.

Webb (2005) states that the integration of ICT with other pedagogical innovations provides even greater potential for enhancement of learning. One clip that many teachers will have used is one of the science experiments completed and filmed at the end of the final Apollo 15 Moon walk. Commander David Scott dropped a hammer and feather, which fell at the same rate. Commander Scott says towards the end of the clip, 'Well, how about that? It looks like Mister [sic] Galileo was correct in his findings!'. The clip can be watched in high definition at: https://nssdc.gsfc.nasa.gov/planetary/lunar/apollo_15_feather_drop.html. In Episode 4 of the BBC Human Universe series, Professor Brian Cox recreates Scott's experiment on a larger scale. Sometimes, using clips like these alongside a practical demonstration can help students to focus in on the phenomena you wish them to experience and replays of the video clip can allow further opportunity for thinking and discussion.

Conceptually difficult content

Protein synthesis is one of the most difficult parts of the current GCSE biology curriculum to teach. Although clearly a very important cellular process, students find it difficult to understand because of its complexity, the involvement of a significant volume of new terminology and its conceptual difficulty. An excellent animation has been produced by the National Human Genome Research Institute, called *Our molecular selves* and available at: https://www.genome.gov/25520211 /online-education-kit-our-molecular-selves/.

Watching short clips through several times whilst answering any questions in between viewings helps students' understanding. Sometimes muting the soundtrack, and the teacher or students using their own words to provide a commentary, can be helpful to consolidate and check on understanding. This website, like many others, allows you to download the transcript. This can obviously be printed or adapted for students and used by them, for example, in the planning of an extended-response question or, more generally, for note-taking or revision.

In physics, an example of a similarly conceptual topic is the relative sizes of objects, from the universe to quarks. An excellent animation to support this learning is found at: https://micro.magnet.fsu.edu/primer/java/scienceopticsu/powersof10/.

Hand-held voting systems

Purpose-made interactive voting systems involve each student using a numbered handset that allows them to vote for answers to questions that have been displayed on the screen. This allows the teacher to see in real time who can answer questions on specific topics and to record that student's progress over time.

Recently, Apps such as Kahoot™ allow students to do much the same as the handsets described above, but on their mobile phones, tablets or laptop computers. Teachers or students can write a quiz and then students answer the questions using their own devices. With Plickers™, students are given a laminated QR code, whose orientation determines their response to a multiple-choice question, and the teacher simply scans the room with his/her own device. This again provides both the students and their teachers with data for formative feedback.

Exciting and inspiring our students

For almost as long as there have been schools in existence, there have been students asking 'Why do we have to learn this?'. If an activity uses media that make learning relevant to them, then they are more likely to engage with it (Webb, 2005). We can do this by choosing examples that they know, such as recent feature films, TV shows or cartoons. For example, the importance of mutations as a driver of evolution is key in the GCSE biology curriculum. Using a recent *X-men*™ trailer in class to introduce this concept captures students' attention.

A word of caution is needed here, because some films illustrate bad examples of science. In the 1994 film *Speed*™ with Keanu Reeves, a bus jumps a 50-foot gap in the freeway! Fortunately, the *Mythbusters*™ TV programme shows that this event couldn't happen, but there are many other bad examples of science in such films and shows. A better example is *Interstellar™* (2014). The *American Journal of Physics* (James *et al*, 2015) published an article praising the accuracy of the science in this film (especially that of wormholes) and suggested that it should be shown to all students in school science lessons. *The Martian™* (2015) is another, more recent, film with much better science within it.

Developing skills as well as knowledge

Being able to balance chemical equations is a fundamental skill. An ICT resource to help students to learn how to balance equations, developed many years ago as part of a Royal Society of Chemistry project entitled *ChemIT*, is still in use in many schools across the UK. It happens to be a Microsoft PowerPoint file, but retains its interactivity by viewing the slides in the 'edit view' instead of the 'slide show view'. An unbalanced symbol equation is visualised using 'molecules' of atoms arranged in linear clusters, which helps students to interpret the chemical formulae correctly. Balancing the equation is then simply a matter of counting coloured atoms and, because the atoms within each 'molecule' are stuck together as movable images, the golden rule of balancing equations (never change the little numbers!) is never broken.

Saving time by not recreating the wheel

Many teachers often start looking for a new resource by typing 'Teaching resource' and then the name of the topic into Google™. In the UK, this often returns results from the *Times Educational Supplement™ (TES)* website. This site has many resources posted by practising teachers. A free signup generates a login, which is required. At the time of writing, a search for a very specific topic such as 'monoclonal antibodies' generated almost 100 resources. Many of these have been created and uploaded by other teachers wishing to collaborate. If using resources created by others (both professional and homemade), it is essential that you check their suitability beforehand.

Making it personal

Sharing personal stories or anecdotes (usually in the form of photos or clips) with our students is another approach to ensuring engagement. Please note that this is different from sharing personal information, such as social media details and addresses. Following a car crash, one of us (Nick) required reconstructive facial surgery, which involved titanium plates being used to support the bones. The images below of the CT scan before and X-ray after the operation helped bring to life lessons about bones protecting vital organs (biology), the unreactive oxide layer on titanium (chemistry), and the use of CT scans and X-rays in medicine (physics). Related photographs, such as the crumple zone and airbag, can be used for other physics lessons!

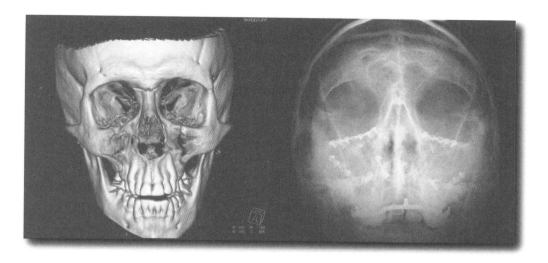

After a particularly exciting chemistry demonstration, Neil had severe burns to his wrist. At the time, students in his Year 9 (age 14) class were filming the reaction using their mobile phones. Needless to say, these clips were online within hours! (See below). Both the videos and photos are useful in teaching about the reactivity series, exothermic reactions, skin, nervous reflexes and healing.

This decision of whether to choose to share information like these examples is of course an entirely personal one. If you are unsure as to whether it is appropriate to share images or other content, it is essential that you speak to your line manager or other relevant person at school.

Collaboration

Haydn and Barton (2006) confirmed that sharing ICT resources in the teaching of science (and history) develops a positive reaction amongst teachers. Collaborating with other members of the department or other teachers in the local area is an important part of teaching, allowing teachers to share ideas and resources. A huge and growing number of teachers are now using Twitter and other social media to drive their own professional networking and continuous professional development. It is encouraging to see how teachers can inspire and support each other, both locally and at a distance. TeachMeets offer a unique opportunity to gather ideas on teaching and learning, often mediated by ICT. If you've not been to one yet, we highly recommend that you do! #ASEChat meets on a Monday evening to discuss and debate science teaching and learning.

One method that can be quickly and easily adopted to promote the sharing of ICT resources and techniques is to have a short 'show and tell' section in department or faculty meetings. Members of staff describe or show other attendees some ICT that they have recently used and explain how it helped in learning and what response they got from their students.

Learning science using ICT: a student's perspective (why and how)

Ofsted, in *Successful Science*, says that *'the very best use of ICT involved students participating actively. Electronic whiteboards used interactively and banks of laptops in laboratories have replaced the need to relocate classes to ICT rooms. Students use laptops for a range of purposes and applications, including:*

➢ *data logging*

➢ *Internet research*

➢ *simulations*

➢ *word processing*

➢ *presentations*

➢ *digital images*

➢ *access to virtual learning environments'* (Ofsted, 2011, p.18).

This said, many schools have limited or no ICT facilities within science laboratories. In this situation, many teachers will take students to specialised IT rooms. Here a number of science programmes, including *Multimedia Science School™* and *Sunflower Science™*, allow students to recreate science experiments on a computer. These simulations have existed now for many years and, in many cases, are very good indeed. We hope their use complements students' learning about practical science (Baggott la Velle *et al*, 2007).

Regarding these programmes, Ofsted reports say that *'Inspectors saw examples of pupils' independent learning at computers using commercial programs, but these were not always sufficiently challenging. Inspectors saw very few examples of pupils using ICT to measure or record the outcomes of practical activities they had done'* (2011, p15). Clearly, ensuring that students are sufficiently challenged by all tasks given to them is an extremely important part of our role as teachers.

The Internet is an amazing resource for students to use in their learning. However, researching on the Internet is difficult, and simply going to an IT room and saying 'Find out about…' is too complex a task for many students. Without scaffolding and, of course, depending upon your context, many students do not possess the skills to find appropriate information to support their learning or stop themselves being distracted. To support this, students can be given a specific website or list of websites to which to restrict themselves. A common example that we use in this regard is *BBC Bitesize* (http://www.bbc.co.uk/education). Furthermore, software such as *Impero Console™* allows teachers to see what websites students are looking at when they are in IT rooms. This has an obvious and important role in helping students stay on task and not be distracted. We find that simply saying 'I can see what websites you are looking at' is usually enough to keep our students on task.

To help support students, teachers can create worksheets in word processing packages such as Microsoft Word, by creating a document that is a 'form' for completion, featuring fields into which comments can be typed with the rest of the text locked. Often, these forms can include drop down menus as well as textboxes for students to type into. These documents can be downloaded from the school shared area, completed by the student during the lessons, and then saved or printed to go into a revision file or exercise book. These worksheets can be written to closely mirror the text that students will encounter on chosen websites, or diverge from it for higher attaining students.

The use of ICT in our lessons can help to scaffold the development of independence for our students (Rogers & Twidle, 2012). Often, after teaching a unit of work, students use an online revision resource such as *BBC Bitesize* to complete a revision homework. It is written by practicing teachers, has age- and awarding body-specific material, only possesses the key information that students need to revise, and offers excellent images, animations and video clips. Many of these have been designed specifically for the website, whilst others come from the enormous archive of science programmes published and produced by the BBC. At the end of each section is a short multi-choice test, which self-marks and provides feedback about any incorrect answers.

The learned societies have websites from which students can access high quality, age-specific resources from which to revise. The Royal Society of Biology runs

Scibermonkey, (see http://www.scibermonkey.org). The Institute of Physics has a similar site, http://www.physics.org/explore.asp and the Royal Society of Chemistry offers *Learn Chemistry* (http://www.rsc.org/learn-chemistry).

Data logging

Data logging involves using sensors and probes to take measurements during practical investigations. Data logging is obviously used extensively in 'real world' science research occurring in universities and the science and technology industries. However, the context in schools is very different. During its 2011 review, Ofsted noted that *'Inspectors saw very few examples of pupils using ICT to measure or record the outcomes of practical activities they had done'* (p.15).

Why is effective data logging used so rarely in science classrooms? Many contexts exist in which data logging can provide genuine learning opportunities, but many teachers still use standard laboratory apparatus, such as thermometers, rather than temperature probes. Simply using a temperature probe to plot a cooling curve of some molten chocolate or wax may not add much value when compared to using a traditional thermometer and plotting a graph by hand (which is a skill that students need to demonstrate in the exams). However, using light gates to investigate acceleration and terminal velocity of an object moving down a slope is an example where the ICT method is mostly superior to using ticker tape machines. Another genuinely advantageous example would be that using data loggers to measure the changes in dissolved oxygen over a 48-hour period in a tank containing some pondweed would extend the traditional practical of counting bubbles per minute to get an idea of photosynthetic rate.

Data logging equipment has traditionally cost a great deal of money, although competition in the market has driven down prices in recent years. Where probes and interfaces are connected to portable computers, there is a real danger of sensors and probes becoming obsolete as the computer hardware and software are upgraded. There is also a need for adequate and enthusiastic technician support.

Students are the experts

Younger people (our students in this case) are more adept at using ICT in many cases than we are. Many teachers make use of this situation to support science teaching and learning by handing over the control to students in their classes.

Examples of how to do this include:

> Asking students to find a resource to share with the rest of the class;
> Making short video clips or animations to cover key concepts;

> ➤ Using science club or specific students to investigate a new technology or resource and help less confident teachers with it. Some schools have a small number of students named 'Digital Explorers' who do this on a more formal basis. Their input is important in deciding which resources are then bought as a school; and

> ➤ Looking to the most competent ICT students first when faced with a technical issue, because they might have the ability and experience to solve it faster than requesting support from an ICT technician.

Many schools now allow students to use their mobile phones in the classroom. The computing power of many of these is incredible. A frequently spoken phrase is that the computing power of one of these devices is more than NASA used to put Neil Armstrong on the Moon. This in itself is surely worthy of discussion in our classrooms!

On a very basic level, students' mobile phones often allow them easy access to the Internet. This means that they can access resources such as clips shown to them in class. There are also a number of amazing Apps with scientific content that can be downloaded. These include:

> ➤ https://www.visiblebody.com for an App that allows limitless views inside the human body;

> ➤ http://www.rsc.org/periodic-table – the Royal Society of Chemistry's Periodic Table; and

> ➤ http://solarsystem.touchpress.com for an App that has amazing images, interactive scenes and videos of the objects in our Solar System.

The biology and physics examples are, sadly, not free, although well worth the price of purchasing.

Other Apps can help students to revise. There are quite a few of these that students will need to pay for. However, the *BBC Bitesize* App is free: (http://www.bbc.co.uk/guides/zgd682p). Again, it is written by practising teachers, is awarding-body specific, contains only the most important information and has short quizzes to test understanding. We highly recommend this.

Mobile phones also allow students to photograph scientific phenomena for themselves. They can take photographs down the microscope to place in their books alongside their microscope drawings and even turn their phone into a microscope by attaching a microscope lens. The Natural History Museum sells one of these, *Phonescope,* for around £10 and many others are available to purchase online (https://www.sciencedaily.com/releases/2016/10/161017130123.htm). Students can also photograph apparatus set-ups, and results (particularly before and after colour changes) that they can come back to later and use in their reports or for discussion.

In conclusion

In summary, using ICT supports teaching and learning in science because it:

> ➢ saves time;

> ➢ makes lessons more interesting and relevant;

> ➢ promotes independence in students' learning;

> ➢ helps students to see that science is not just taught in classrooms but is in everyday life;

> ➢ fosters a more collaborative way of working; and

> deepens students' understanding of science concepts.

As well as this, it is also one of the more creative and enjoyable aspects of science teaching.

Further reading

EEF (2017) *Teaching and Learning Toolkit – Digital technology*. Available at: https://educationendowmentfoundation.org.uk/resources/teaching-learning-toolkit/digital-technology/

Websites

http://www.youtube.com
http://www.sciberbrain.org
https://nssdc.gsfc.nasa.gov/planetary/lunar/apollo_15_feather_drop.html
http://www.bbc.co.uk/programmes/p02985m0
https://www.brainpop.co.uk
https://www.genome.gov/25520211/online-education-kit-our-molecular-selves/
https://micro.magnet.fsu.edu/primer/java/scienceopticsu/powersof10/
https://www.youtube.com/watch?v=dbQFfOTsr6c
http://www.bbc.co.uk/education
http://www.scibermonkey.org
http://www.physics.org/explore.asp
http://www.rsc.org/learn-chemistry

https://www.visiblebody.com
http://www.rsc.org/periodic-table
http://solarsystem.touchpress.com
http://www.bbc.co.uk/guides/zgd682p
https://www.sciencedaily.com/releases/2016/10/161017130123.htm

References

Baggott la Velle, L., Wishart, J., McFarlane, A., Brawn, R. & John, P. (2007) 'Teaching and learning within the subject culture of secondary school science', *Research in Science and Technological Education*, **25,** (3), 339–349

EEF (2017) *Teaching and Learning Toolkit – Digital technology*. Available at: https://educationendowmentfoundation.org.uk/resources/teaching-learning-toolkit/digital-technology/

Haydn, T. & Barton, R. (2008) 'First do no harm: Factors influencing teachers' ability and willingness to use ICT in their subject teaching', *Computers and Education*, **51,** (1), 439–447

James, O., von Tunzelmann, E. & Franklin, P. (2015) 'Visualising *Interstellar*'s Wormhole', *American Journal of Physics*, **83,** (6), 486–499

McFarlane, A. & Sakellariou, S. (2002) 'The role of ICT in science education', *Cambridge Journal of Education*, **32,** (2), 219–232

Ofsted (2011) *Successful Science: an evaluation of science education in England 2007-2010*. No. 100034. Available at: https://www.gov.uk/government/publications/successful-science-strengthes-and-weaknesses-of-school-science-teaching

Rogers, L. & Twidle, J. (2013) 'A pedagogical framework for developing innovative science teachers with ICT', *Research in Science and Technological Education*, **31,** (3), 227–251

Webb, M. (2005) 'Affordances of ICT in science learning: implications for an integrated pedagogy', *International Journal of Science Education*, **27,** (6), 705–735

CHAPTER 16

Language and talk in science education

Richard Taylor

Learning science requires our students to learn both the ideas in science and the language that scientists use to talk about them. Much has been written about language in science and, in this chapter, the author presents an overview of the research into why learning scientific language can be challenging, and suggests strategies to help support students' use of scientific language in talk, reading and writing.

The importance of language in science education

One of the principles of science education is that schools *'should aim systematically to develop and sustain learners' curiosity about the world, enjoyment of scientific activity and understanding of how natural phenomena can be explained'* (Harlen, 2010, p.5), and much of science education can be viewed as learning to explain the world around us. However, children come to science lessons having already developed explanations of the natural world (Osborne, Bell & Gilbert, 1983), as already discussed in Chapters 1 and 4.

It has been suggested that the process of learning science might mirror the scientific process, in which children's naïve theories or explanations are falsified using experimental evidence and then replaced with more logical scientific theories that are consistent with this evidence (e.g. Posner, Strike, Hewson & Gertzog, 1982). However, Joan Solomon highlights one important difference between the process of learning science and 'doing' science by suggesting that we pay attention to the way in which expert scientists talk (Solomon, 1985, 1994). She points out that it would not be unusual to hear a physicist making statements that apply physical theories incorrectly. For instance, a physicist might explain his/her exhaustion after a long run by saying *'Phew, I've used up all my energy'*, even though s/he would be aware of the illogical nature of this statement given his/her knowledge of the principle of conservation of energy. A more correct statement would be *'I've transferred a large amount of chemical energy from my food to kinetic and thermal energy during that run. So I need to eat something to replace this chemical energy. I will then feel less exhausted'*. Clearly, physicists do not talk in this way! However, you would not expect a physicist to use a defunct scientific theory to explain a natural phenomenon.

So why do expert scientists – and indeed almost everyone with some scientific knowledge – persist in using ideas about energy that are clearly wrong and, more importantly, that they know are wrong? The answer to this question lies in the dictionary definitions of energy:

1. *Strength and vitality required for sustained physical or mental activity.*
2. *The property of matter and radiation, which is manifest as a capacity to perform work* (OED Online).

These multiple definitions illustrate that the meaning of language depends on the context in which it is being used and, hence, there is a language of science, which can overlap with, but is often distinct from, everyday language. Recognising this distinction between scientific and everyday language is important, because learning science is a social process, as Jay Lemke points out:

'*Reading, writing, hearing and especially talking science are a larger part of what professional scientists do. Along with some time spent in practical work, they are most of what science teachers and pupils do*' (Foreword in Wellington & Osborne, 2001).

Lemke is highlighting that – whilst scientists do, of course, learn through experiments – they primarily develop new ideas and theories through use of written and spoken language. Similarly, science teachers use written and spoken language to help their students learn scientific ideas. Hence, despite the appeal of the analogy between a student learning a new scientific idea through experimentation and a scientist accepting or rejecting a theory in the face of new experimental evidence, we cannot escape the reality that, most of the time, students' understanding of science is talked, read or written into existence (Lemke, 1990; Mortimer & Scott, 2003; Sutton, 1992; Wellington & Osborne, 2001).

Given this importance of language in learning and the subject-specific nature of language used in science, it is clear that students need to learn scientific language to understand the ideas that are being communicated. Moreover, there is evidence that learning to talk scientifically helps one to learn to think scientifically (Carlisle, Fleming & Gudbrandsen, 2000). But what are some of the challenges that students face when learning scientific language, and how might teachers use spoken and written language to support students' learning?

Challenges when learning scientific language
Scientific words
Children are prolific word learners. The average child will have vocabulary of about 60,000 words by the time s/he has left secondary school (Miller, 1996;

Pinker, 1994). To achieve this feat, children begin to learn words from around 12 months old and then learn them at an increasing rate as they age, with suggestions that the average 8-10 year-old will be learning words at a rate of approximately 12 words per day (Bloom, 2000). If all these words are learnt at school, across a five-lesson day, it would suggest that an average Year 7 (age 12) student has the capacity to learn approximately two or three new words per lesson.

Now, consider an average Year 7 student who has been asked to read the following explanation of how she smells the perfume or deodorant she is wearing:

'Perfume **particles evaporate** from your skin. The **particles** move around **randomly**. They mix with the air. As the perfume **particles** spread out, some enter your nose. Your nose detects the smell. The **random** moving and mixing of **particles** is called **diffusion**' (Gardom Hulme, Locke, Reynolds & Chandler-Grevatt, 2013 p.70).

The words in bold are likely to be new – or new in terms of their scientific meaning – to this student: five unfamiliar words in one passage, or just five minutes, or double their expected capacity for a lesson. This example illustrates that it is the density of scientific or technical terms that requires students to learn words at a greater rate than expected for their age. This is one reason why science is challenging to learn (Wellington & Osborne, 2001). Hence, Maskill (1988) suggests that, as science teachers, we should remove or simplify these technical words where possible. However, Wellington and Osborne (2001) argue that this approach is inadvisable given that 'learning to use the language of science is fundamental to learning science' (p.6).

So why does learning scientific language help with learning science? Halliday and Martin (1993) provide one answer to this question. Their research examined how eminent scientists, including Isaac Newton, wrote about their ideas, and they identified a technique called nominalisation. Nominalisation involves a complex process or concept being given a name to 'package' it into what Halliday and Martin (1993) call a 'single semiotic entity' – usually a word. For example, 'the bending of light as it crosses a boundary between material with different densities' becomes 'refraction'. Halliday and Martin (1993) argue that this process of nominalisation allows scientists to communicate their ideas more elegantly, succinctly and, perhaps, more persuasively.

Halliday and Martin's arguments have several implications for science education. First, high-density technical terminology in scientific language is not just scientists trying to sound clever. This frequent use of terminology is necessary for scientists to communicate their ideas effectively and may play a role in enabling scientists to effectively apply and develop these ideas. Hence, learning technical terms (and many of them) is a necessity when learning scientific language.

Secondly, nominalisation generates technical terms that represent complex processes and concepts. They also represent ideas that are abstract and cannot be understood in isolation. This richness and complexity of knowledge around some scientific words implies that students need time to learn the meaning of technical terms, that they can find this process difficult and are likely to need assistance (Best, Dockrell & Braisby, 2006; Dockrell, Braisby & Best, 2007; Wellington & Osborne, 2001).

Dual-meaning terms

Dual-meaning terms have subtly different meanings in everyday and scientific contexts (Itza-Ortiz, Rebello, Zollman & Rodriguez-Achach, 2003). For instance, *acceleration* typically means 'getting faster' in an everyday situation, rather than 'the rate of change of velocity' in a scientific context. The presence of an existing meaning makes it challenging for students to learn the new scientific meaning (Jasien, 2010, 2011), and can lead to miscommunication due to a lack of shared meaning between teacher and student. For instance, try asking Year 7 students (or perhaps even Year 11, age 16) if they think a falling object accelerates when it hits the floor. Hence, as science teachers, we need to be aware of the problems that these dual-meaning terms create for students and should try to mitigate these problems by highlighting their subtle, context-dependent differences in meaning.

Cassels and Johnstone (1985) found that the understanding of a large number of secondary school students of what they called non-technical science terms (e.g. 'emit', 'composition', 'negligible', 'contrast', 'fundamental') was also poor. Similarly, Gardner (1977) found that students did not understand some logical connectives (e.g. hence, moreover), which are often used for constructing explanations in science. Again, these studies emphasise that, as science teachers, we can often unknowingly speak in a language that our students do not understand and we need to be aware of this.

Supporting the learning of scientific language

Social-constructivist models of learning (e.g. Vygotsky, Bruner) often describe language as a tool for reasoning, and have played a major role in researchers and teachers recognising the role of language in learning science (e.g. Leach & Scott, 2003). One analogy suggested by Wertsch (1985) identifies different types of language as different toolkits for problem-solving. However, this analogy is limited, as you would not expect real tools to change simply as a result of using them. Neither would you expect new tools to be added to your toolkit. However, this is not the case with language. Our understanding of language often changes when we carry out activities – talking, reading and writing – that involve language.

For instance, scientific words often represent abstract concepts. Hence, learning the concepts represented by these words, thus learning the meaning of these words, cannot be achieved simply through sensation (Bloom, 2000). For example, a person might learn the concept of a dog and the meaning of this word by playing with several of 'man's best friends'. However, as Bloom (2000) argues, the same mechanism cannot apply to words that represent abstract concepts, such as 'mortgage' or 'atom', because we cannot interact with these concepts in the same way as with a concrete concept, such as a dog. Instead, these words for abstract concepts are learned through linguistic context, which is provided by talking, reading and writing. Hence, this section looks at strategies that teachers can use to engage students in these activities and specifically to promote development of scientific language. Most of these strategies are taken from Wellington and Osborne (2001).

Talk as a mechanism for learning scientific language and developing reasoning

Talk is both a powerful catalyst for learning scientific ideas, and a very natural and familiar way of communicating ideas and giving explanations. However, as science teachers, our familiarity with talk should not lead us to ignore the complexity of the different types of talk used in the classroom, or to fail to recognise that sometimes students need to be taught how to use these different kinds of talk.

Teacher-student talk

Science teachers often face a difficult dilemma (Wellington & Osborne, 2001). On the one hand, they need to use the process of learning scientific ideas to model the nature of science as a genuine process of discovery and fertile ground for new thinking. However, they also need to present much of the science curriculum as an accepted body of knowledge. The strategies that science teachers use to encourage different types of talk reflect the balance they need to strike to achieve these goals. These strategies and the types of talk they promote are illustrated by the examples of dialogue below, which are taken from a detailed study of talk within the classroom carried out by Mortimer and Scott (2003). Each of these extracts of dialogue have been taken from a series of lessons in which an experienced teacher is developing 13-14 year-old students' – with a wide range of prior attainment – understanding of rusting. The analysis below summarises Mortimer and Scott's (2003) extensive discussion from Chapter 4 of their book:

Episode	Dialogue	Communicative approach	Pattern of interaction
1	The teacher has given each student an iron nail and asked them to put the nail in a place where they think it will rust. Teacher: [...] What do you think it was – thinking about the place – that made your nail go rusty? *Several students list moisture and damp as causes of rust.* Teacher: Moisture. Damp, moisture. Anything else? Gavin? Gavin: I put mine in some mud. Teacher: What was it about that mud that you think made yours go rusty? Gavin: Cos it were all wet and boggy. Teacher: Wet – so it was wet again.	Interactive / dialogic.	Initiation then response then feedback then response etc... (I-R-F-R-F)
2	Teacher: So in fact everyone's got their hand up, telling me that with air and water the nail has gone very rusty. Right – now then. Is that telling us something very important, d'you think? Have we narrowed this information down any more? Dawn? Dawn: Well, it means that, means, er...you have to have them both together for the nail to go rusty. Teacher: Right. I think that is an excellent point – and I think it's an excellent way of saying it too. [...]	Interactive / authoritative.	Initiation then response then evaluation (I-R-E)
3	Teacher: Let's just think back again. At the start, you were suggesting that it was cold, it was warm, it was dark, it was light, it was acids, or it was – water and air. All those things caused rust. That's what we started off thinking. And what we've done now – we've come to the point where you've decided and you've proved in fact that it's *just two* things with the iron.	Non-interactive / authoritative.	No interaction.

In Episode 1, the teacher encourages students to explore their ideas about what causes rusting by using a communicative approach that is interactive and dialogic: interactive in the sense that the students are expected to contribute to the discussion and dialogic because the teacher is clearly interested in understanding students' explanations of what causes rusting. For instance, when Gavin identifies mud as a cause of rusting, the teacher asks him to clarify why he believes that mud causes rusting, rather than simply ruling out mud as a cause. This response signals that all perspectives in this exchange are equally important and the scientific perspective is not prioritised. Moreover, this response guides the dialogue and the development of understanding, rather than evaluates and tests Gavin's scientific knowledge. Hence, this exchange between Gavin and the teacher is also an example of the I-R-F-R-F (initiation – response – feedback) pattern of interaction. Mortimer and Scott (2003) argue that this pattern of interaction effectively establishes the interactive/dialogic communicative approach, which encourages dialogue that models scientific inquiry.

Episode 2 illustrates a mid-point in a continuum between the transmission of science as an accepted body of knowledge and the modelling of science as a process of inquiry. Mortimer and Scott (2003) identify dialogue that serves this purpose with an interactive/authoritative communicative approach. The teacher initiates the discussion with a question to encourage students to share their ideas and place them at the fore, and then evaluates the students' responses against accepted scientific knowledge, rather than simply giving feedback to encourage the students to expand upon their ideas. Hence, this pattern of interaction is an example of I-R-E (initiation – response – evaluation) interaction (Mehan, 1979). Mortimer and Scott (2003) argue that this type of interaction allows teachers to work and shape students' ideas, to encourage them to adopt scientific notions by promoting and emphasising the scientific perspective.

In Episode 3, the teacher stresses the accepted scientific explanation of rusting. To achieve this emphasis, the teacher uses a non-interactive/authoritative communicative approach. There is deliberately no opportunity for questions, either to or from the students. This lack of interaction implies that the teacher is 'transmitting' rather than looking to 'share' or 'shape' ideas (Barnes, 1973). Hence, the teacher creates the impression that she is now the authority on the knowledge and, in this example as is often the case, this authority is used to emphasise the scientific perspective and the 'correctness' of this knowledge.

Mortimer and Scott (2003) stress that all these types of dialogue play a role in introducing students to the scientific story, or the way of thinking about the natural world that is characterised by the use of scientific language (Ogborn, Kress, Martins & McGillicuddy, 1996). Hence, using the types of dialogue is integral to developing scientific language. Moreover, Mortimer and Scott (2003) suggest that there is a rhythm to how these types of dialogue are used, which could constitute a pedagogical approach to developing an understanding of language. First, an interactive/dialogic 'opening-up' phase of dialogue exposes students' current understanding of scientific ideas and the language they associate with these ideas. Then, an interactive/authoritative 'working-on' phase of dialogue shapes students' initial ideas, exposed in the opening-up phase, to develop scientific thinking and language. Finally, a non-interactive authoritative 'shutting-down' phase ends discussion and stresses the scientific point of view and the meaning of any relevant scientific language. This approach might take place over the course of a lesson, or over a sequence of lessons.

Student-student talk

As discussed earlier, new ideas in science are developed as much by discussion as experimentation. But how much of a science lesson is spent in discussion? For example, Newton et al (1999) found that, in their observation of 39 science lessons, less than 5% of lesson time was dedicated to group discussion between peers, as opposed to 30% of the time in teacher-led discussion. Moreover, less than 2% of teacher-student discussion was genuine discussion, rather than questioning to assess students' scientific knowledge. This is perhaps understandable, given the different aims of discussion identified by Mortimer and Scott (2003), but still most questions in science lessons are asked by a teacher, not a student, and a teacher tightly controls the direction of any discussion.

There are several reasons why more lesson time should be devoted to opportunities for students to openly discuss science with their peers. Firstly, students are more likely to openly discuss and test out ideas in the relatively low risk environment of a small group of friends, rather than the high risk, high stakes environment of a whole class discussion hosted by a teacher (Wellington & Osborne, 2001). Moreover, a study by Rowe (1974) showed that students are typically given only a few seconds to respond to a question in a teacher-student discussion. As Wellington and Osborne (2001) point out, this sort of thinking time is hardly likely to be conducive to deep-thinking and considered responses. Secondly, once an environment in which students are willing and are given time to engage in discussion is established, several empirical studies (see Mercer (2008) for a review) cite evidence that suggests this sustained dialogue between peers helps students to collectively solve problems and promotes individual learning of the concepts required to think scientifically (Howe, Tolmie & Rodgers, 1992). Moreover, this dialogue is most productive when students present their own ideas and are asked to justify them (Howe *et al*, 2007). Listed below are a number of activities that Wellington and Osborne (2001) suggest can be used to stimulate discussion amongst small groups of students:

➢ Concept mapping;

➢ Discussion of instances;

➢ Critical reasoning; and

➢ Generating questions.

Learning to read and write scientifically

Wellington and Osborne (2001) suggest that reading and writing are avoided in classroom because students find scientific language difficult and, when they read and write, they are required to engage with this language. However, engaging with scientific language is exactly what students need to do to learn this language. This is discussed in more detail in Chapter 17.

Strategies to encourage scientific reading

For reading in science to be useful, students must reflect on what they are reading (Lunzar & Gardner, 1979). Reflective reading requires active engagement with a text rather than just skimming over the words. Unfortunately, scientific text often lacks a narrative, or storyline, with which to naturally engage students and hold their attention (Wellington & Osborne, 2001). Hence, encouraging students to read scientific text can often involve providing a purpose to sustain engagement and supporting students in achieving the goal that the purpose provides (Davies & Greene, 1984). Strategies designed to provide this purpose are summarised in Appendix 1.

Strategies to encourage scientific writing

The use of writing frames can help students to become familiar with the aims of different genres of scientific writing and scaffold their attempts to write within these genres (Wray & Lewis, 1997). These genres and some examples of appropriate writing frames are summarised in Appendix 2.

In conclusion

As science teachers, we confidently use words like *particle*, *evaporate*, *diffusion* and *random*. These words form part of the language of science, which can overlap, but is often distinct from, the language we use in an everyday context. We speak this language fluently, but our students do not. We use scientific language to think about and articulate our explanations of the natural world and engage in talking, reading and writing about science, which is primarily how we have learnt science. The importance of language in learning science makes developing our students' language skills an important goal of science education. This chapter has set out a number of strategies that can be used to achieve this goal.

Strategy	How to carry out the activity (APPENDIX 1)
General versus specific instructions	This strategy simply encourages the use of specific rather than general instructions to make the purpose of any reading activity clear to students. Some examples of specific rather than general instructions are given below. **GENERAL INSTRUCTIONS** Read this chapter about electric motors for homework. Read and make notes about the eye using pages 131 to 132. **SPECIFIC INSTRUCTIONS** Underline all the words and phrases that refer to parts of an electric motor. Use pages 131 and 132 to label the parts of the eye and explain what each part does.
Directed Activities for Science Texts (DARTs)	DARTs encourage students to focus on important parts of a selected text and to reflect on its content by looking for specific information. Students are often asked to mark and label text to indicate the location of the information or record and reconstruct the information when they have found it. **MARKING AND LABELLING** Underlining/marking – students search for targets in text (e.g. words or sentences). Labelling – students label parts of a text with labels provided for them (e.g. description of how rocks are weathered). Segmenting – students break the text down into pieces, or units of information, summarise the information each segment gives. **RECORDING AND RECONSTRUCTING** Students construct a diagram to visualise the content and flow of the text – a flow diagram, mind-map or branching tree etc... Key points/summary – limit students to only one or two sentences or bullet points to summarise the text. Limiting the amount students can write encourages them to pick out key messages.

Genre	Writing Frame	(APPENDIX 2)
Reports Including reports that: • classify, • describe functions and process, • list properties.	**Example 1** The animal I am describing is… It normally lives… It feeds on… During the day, it can be seen doing…	**Example 2** The part of the body I am describing is… It consists of… The purpose of each part is… If you drew it, it would look like…
Explanations	I want to explain why… solids cannot be compressed. This is because… the particles in a solid are touching.	
Experimental accounts	**Aims** • What is the purpose? • Why are we doing this? • What is our prediction and hypothesis? **Method** • What is the 'recipe' for doing this experiment? • What are the instructions? • Would a diagram help to explain what we did? **Results** • How should we display the results – table, bar chart, line graph? **Conclusions** • What do our results show? • What happened in the experiment? • Can you describe a relationship? • Can we trust our results? • Did what happened in the experiment match what we expected might happen? • Do our results agree with our prediction and support our hypothesis? • Does our experiment change any ideas we have about science?	
Argument	There is a lot of discussion about… The people who agree with this idea claim that… They also argue… However, there are also strong arguments or evidence against this view. These arguments are… After looking at these different points of view, I think that…	

Further reading

Lemke, J.K. (1990) *Talking science.* Norwood, NJ: Ablex

Mortimer, E.F. & Scott, P.H. (2003) *Meaning making in secondary science classrooms.* Maidenhead: Open University Press

Sutton, C.R. (1992) *Words, science and learning.* Milton Keynes: Open University Press

Wellington, J.J. & Osborne, J. (2001) *Language and literacy in science education.* Buckingham: Open University Press

References

Barnes, D. (1973) *Language in the classroom.* Milton Keynes: Open University Press

Best, R.M., Dockrell, J.E. & Braisby, N.R. (2006) 'Lexical acquisition in elementary science classe', *Journal of Educational Psychology,* **98,** (4), 824–838

Bloom, P. (2000) *How children learn the meanings of words.* Cambridge, Mass: MIT Press

Carlisle, J.F., Fleming, J.E. & Gudbrandsen, B. (2000) 'Incidental word learning in science classes', *Contemporary Educational Psychology,* (25), 184–211

Cassels, J.R.T. & Johnstone, A.H. (1985) *Words that matter in science – a report of a research exercise.* London: Royal Society of Chemistry

Davies, F. & Greene, T. (1984) *Reading for learning in the sciences.* Edinburgh: Oliver and Boyd

Dockrell, J.E., Braisby, N.R. & Best, R.M. (2007) 'Children's acquisition of science terms: Simple exposure is insufficient', *Learning & Instruction,* (17), 577–594

Gardom Hulme, P., Locke, J., Reynolds, H. & Chandler-Grevatt, A. (2013) *Activate 1.* Oxford: Oxford University Press

Halliday, M.A.K. & Martin, J.R. (1993) *Writing science: Literacy and discursive power.* London: Falmer Press

Harlen, W. (Ed.) (2010) *Principles and big ideas of science education.* Hatfield: The Association for Science Education. Available at http://www.ase.org.uk/resources/big-ideas/

Howe, C., Tolmie, A. & Rodgers, C. (1992) 'The acquisition of conceptual knowledge in science by primary school children: Group interaction and the understanding of motion down an incline', *British Journal of Developmental Psychology,* **10,** (2), 113–130

Howe, C., Tolmie, A., Thurston, A., Topping, K., Christie, D., Livingston, K. & Donaldson, C. (2007) 'Group work in elementary science: Towards organisational principles for supporting pupil learning', *Learning and Instruction*, **17,** (5), 549–563

Itza-Ortiz, S.F., Rebello, N.S., Zollman, D.A. & Rodriguez-Achach, M. (2003) 'The vocabulary of introductory physics and its implications for learning physics', *The Physics Teacher*, (41)

Jasien, P.G. (2010) 'You said "neutral", but what do you mean?', *Journal of Chemical Education*, **87,** (1), 33–34

Jasien, P.G. (2011) 'What do you mean that "strong" doesn't mean "powerful"?', *Journal of Chemical Education*, **88,** (9), 1247–1249

Leach, J. & Scott, P. (2003) 'Individual and sociocultural views of learning in science education', *Science and Education*, **12,** (1), 91–113

Lemke, J.K. (1990) *Talking Science*. Norwood, NJ: Ablex

Lunzar, E. & Gardner, K. (Eds.) (1979) *The effective use of reading*. London: Heinemann

Maskill, R. (1988) 'Logical language, natural strategies and the teaching of science', *International Journal of Science Education*, **10,** (5), 485–495

Mehan, H. (1979) *Learning lessons: Social organization in the classroom*. Cambridge, MA: Harvard University Press

Mercer, N. (2008) 'Talk and the development of reasoning and understanding', *Human Development*, (51), 90–100

Miller, G.A. (1996) *The Science of Words*. New York: Freeman

Mortimer, E.F. & Scott, P.H. (2003) *Meaning making in secondary science classrooms*. Maidenhead: Open University Press

Ogborn, J., Kress, G., Martins, I. & McGillicuddy, K. (1996) *Explaining science in the classroom*. Buckingham: Open University Press

Osborne, R.J., Bell, B.F. & Gilbert, J.K. (1983) 'Science teaching and children's views of the world', *European Journal of Science Education*, **5,** (1), 1–14

Pinker, S. (1994) *The Language Instinct*. New York: Harper Collins

Posner, G.J., Strike, K.A., Hewson, P.W. & Gertzog, W.A. (1982) 'Accommodation of a scientific conception: Toward a theory of conceptual change', *Science Education*, (66), 211–227

Rowe, M.B. (1974) 'Relation of wait-time and rewards to the development of language, logic, and fate control: Part ii-rewards', *Journal of Research in Science Teaching*, **11,** (4), 291–308

Solomon, J. (1985) 'Teaching the conservation of energy', *Physics Education*, (20), 165–170

Solomon, J. (1994) 'The rise and fall of constructivism', *Studies in Science Education*, (23), 1–19

Sutton, C.R. (1992) *Words, science and learning.* Milton Keynes: Open University Press

Wellington, J.J. & Osborne, J. (2001) *Language and literacy in science education.* Buckingham: Open University Press

Wertsch, J.V. (1985) *Vygotsky and the social formation of the mind.* Cambridge, Mass: Harvard University Press

Wray, D. & Lewis, M. (1997) *Extending literacy: Children reading and writing non-fiction.* London: Routledge

CHAPTER 17

Literacy in science: a platform for learning

Billy McClune

*As science teachers, we have a responsibility to develop students'
communication skills and at the heart of this is literacy. This chapter explores
the rich resource of science in the news as a context for learning and the
potential pathway for the development of literacy in science. It builds on the
discussion of language and talk in Chapter 16. Exemplar news reports illustrate
some practical approaches to working with science in the news.*

Introduction

*'The effort and time that teachers give to strengthening students' language
skills through science is an investment not an expense'* (Their, 2002, p.35).

Teachers who invest in the development of language skills are likely to do so
because they view language skills and literacy capability as the foundational
elements of the platform from which science is explored. Language is the
agency through which science is understood and communicated.

Consequently, many teachers ensure that language-rich activities are integral
and prominent among their teaching approaches to support learning in
science. Through reading, writing and talk, young people can be equipped to
engage meaningfully with the established science that surrounds them today
and the yet-to-be-discovered science that they will encounter tomorrow.

Scientific literacy is a long-term goal of science education. It is a competence
that develops gradually. It flourishes in environments that encourage the
coming together of knowledge, skills, attitudes and ways of thinking and
questioning in science that are characteristic of science teaching at its best.
The important thread that links these elements is language. Viewed from the
perspective of science concepts and skills, the student who is becoming
increasingly scientifically literate is learning to use the language of science to
do science. Participating in science involves engaging in the discourse and
practice of science, using terminology, reasoning in a scientific context, making
observations, handling data and constructing meaningful links between
experience and knowledge.

Wellington and Osborne (2001) effectively make the case for the centrality of language in science literacy. They have brought together key ideas from different literacy strands to highlight issues and to offer practical guidance. Others have added to this by focusing on specific aspects of literacy, such as reading (Hines *et al*, 2010). From the perspective of language and communication, the student who is becoming increasingly scientifically literate is learning to use language *in* science. Language is used to draw conclusions, formulate claims and structure evidence. Students are using language effectively, not only to communicate their grasp of knowledge and ideas, but also to construct and shape that knowledge. The literary elements of writing and reading, the spoken language elements of talking and listening and the presentational elements of representing and viewing are all involved in learning science. The view of literacy underpinning this chapter embraces all these literary, auditory, oral and representational elements as essential epistemic instruments.

One goal of science education is that students should achieve a level of 'subject literacy' that would enable them to tackle, with confidence, science-related issues that affect them as individuals, communities and as a society. We want young people to be able to understand, and use appropriately, science-informed reasoning in their day-to-day living. Scientific literacy is generally accepted to integrate:

➢ Subject knowledge – in the science curriculum built around the 'big ideas' in science;

➢ An appreciation of the nature of science that includes the ways of doing and ways of thinking in science; and

➢ The capacity and aptitude to engage critically with science as it impacts on the lives of individuals, communities and society.

Big ideas and science news

Many aspects of daily life are shaped by science knowledge and its application. It influences work, leisure and lifestyle. The big ideas of science (Harlen, 2010) underpin many of our day-to-day experiences. They help us to understand, explain and therefore engage meaningfully with the world around us. However, meaningful engagement as capable, informed, scientifically-literate individuals, who are able to contribute to decisions that affect individuals and their communities, begins with science in school. The science teacher's role is to equip individuals for a lifelong journey of encounters with science.

Science issues in the world around us crop up regularly in the news. There are new discoveries, groundbreaking science applications and socio-scientific dilemmas that are likely to be aired in the news media, posted on the web and shared through social media and personal communications. By engaging with science-based news, teachers invite these issues into the classroom. Through doing this, they can use these language-rich resources to explore concepts and applications of science and also promote fundamental literacy. In equipping students with the tools to engage with the emerging science issues that crop up in the world around them, teachers are nurturing the literacy element in scientific literacy.

Literacy in policy and practice

In many parts of the world, scientific literacy is recognised as a curriculum goal. However the term is imprecise, as Jenkins (2010 p.27) notes: 'It is a useful overarching goal capable of sustaining a variety of meanings'. Advocacy for reasoning in science – viewing science as a way of knowing – is gaining support (Norris & Phillips, 2003) and has led to reporting on scientific literacy as an element in international comparisons through the Programme for International Student Assessment (PISA). This has raised the profile of scientific literacy internationally. In some curricular documents, there is specific emphasis on increasing scientific literacy or developing a scientifically-literate citizen (CCEA, 2012; Education Scotland, 2010); in others, elements of becoming scientifically literate are included under the theme 'working scientifically' (National Curriculum England and Wales, 2005).

Policy documents that inform curricula highlight the importance of placing science in settings that are familiar to students and that encourage contextualised teaching. In writing about the principles that underpin science education, Harlen (2010, p.7) includes 'enabling individuals to take an informed part in decisions...that affect their own wellbeing and the wellbeing of society and the environment'. For science, this could be enhanced through cross-curricular collaboration. While collaboration with the English Department may be an obvious starting point, other collaborators are also useful, such as with geography, history, art, media and technology.

Literacy and science communication in everyday life

Decision-making about lifestyle may be influenced by science news media. This may guide actions when people engage with environmental or health issues that affect them personally, or their community or society. An individual's ability to engage critically with media reports with a science component is one indicator of his/her level of scientific literacy. Therefore, media-based science communication generally, and news in particular, is an appropriate context for addressing and developing literacy in science.

News-based science reporting, however, is not like science in a textbook. It is a complex entity with embedded values and viewpoints. Journalists adhere to certain established conventions that shape the content and presentation of news across many different platforms. News is also 'tweeted', 'liked' and shared between individuals and groups with similar interests and outlooks. To access it effectively and evaluate it thoughtfully requires a level of literacy that science education may not currently provide.

The nature, scope and potential influence of science communication is such that it could be argued that school science has a responsibility to:

> Encourage students to pay attention to news so as to be aware of its impact and to recognise its potential as a source of ongoing learning;

> Explore science news as a platform for learning science;

> Promote science news as a language-rich environment and a means to develop literacy;

> Equip students to read science news critically; and

> Consider the opportunity that science news offers for interdisciplinary learning.

Science-related news reports can provide an opportunity for developing an aptitude and ability to engage with science in the media. In addition, the language-rich resources provide opportunities for learning about reading, writing and talking. With news media, students can both learn *about* and also learn *through* language in science. In this section we will explore how news items can be exploited to achieve these literacy goals, together with learning in science.

A number of theoretical constructs underpin thinking and practice in relation to the inclusion of science-based media reports as an element of the school science curriculum. These are that:

> science literacy should equip young people to engage with science in the world beyond the classroom;

> science literacy should develop students' capability for critical reading of science-based media;

> science reported in the media has unique and challenging characteristics and requires particular attention in the science classroom; and

> the learning intentions and pedagogies associated with effective use of media-reported science have implications for interdisciplinary approaches to teaching.

Literacy as an intentional activity

For each new science topic, language is foundational in initiating the learner into the topic vocabulary and discourse. As students listen, read, write, visualise or talk about the targeted concept, they have the potential to develop their literacy skills. The science teacher who is alert to these possibilities can exploit the literacy opportunities as they arise naturally, while intentionally choosing activities for their influence on literacy development and contribution to science learning. Introducing news media is one of those intentional choices. An example of how this might be approached is given in Figure 1:

Figure 1: Traffic noise reduces wild owls' foraging efficiency

Science daily	September 27, 2016	
[1]	Traffic noise reduces wild owls' foraging efficiency Source: Hokkaido University	[1] Thinking & Talking
[2]	Traffic noise reduces the hunting ability of wild owls by up to 89 percent, a team of researchers has found. The team's world-first study examined how different levels of noise affect nighttime hunters, which use their sense of hearing to find prey in the darkness.	
[3]	As levels of traffic around the world increase, many people are concerned that traffic noise will badly affect wild animals, A team of researchers from Hokkaido University used an artificial prey rustling sound to attract owls and help them observe the birds' night time hunting activity.	[2,3,6] Media insights & reasoning
[4]	The team conducted experiments at 103 locations in Japan where owls overwinter. In the experiments, researchers studied the activities of 78 owls while playing back different levels of recorded traffic noise and a constant level of an artificial prey rustling sound.	
[5]	The experiments showed that the rate of owls detecting the artificial sound reduced by 17 percent even at the lowest traffic noise level of 40dB (equivalent to the noise level of a quiet residential area), while the rate dropped by 89 percent at the highest noise level of 80dB (similar to the noise level inside a train).	[4,5] Reading & Questioning
[6]	"Changes in the way owls hunt can alter the balance between prey and predators, and possibly have a negative effect on the entire ecosystem," said Futoshi Nakamura, one of the coauthors at Hokkaido University.	[6] Talking, Listening & Reasoning
[7]	We need to know more about how the noise affects the birds in order to develop measures to reduce the impact of traffic noise on animals that depend on sound to hunt their prey.	[7] Multimodal communication

The news report (*Science Daily*, 2016) featured in Figure 1 contextualises noise pollution and highlights an example of environmental impact. The subject matter links well to the topic of sound and lends itself to numerous practical activities and student-directed inquiry involving sound level measurement and materials with sound-insulating properties to reduce noise pollution.

A teacher wishing to exploit the literacy opportunities may begin with just the news headline to introduce discussion about the topic (see Figure 1 [1]). Students might be asked to think about what lies behind the headline. Why could traffic noise be a problem? In an open-ended approach, students might be invited to pose their own questions stimulated by the headline. They could search the text for specific information. They could be asked to find and comment on specific elements of text. Where was the study conducted? How certain are scientists about the effect of traffic noise? What are the sources of the report? How was the report structured so as to communicate science effectively?

Much is to be gained by giving students opportunities to explore their ideas through talking and listening, though careful thought needs to be given to the management of these activities. Students may need some time to think before talking begins. It is important to have a space, possibly created through a 'classroom code for talking and listening', where each voice can be heard and all ideas can be valued. Some mechanism to filter the ideas and manage the outcome for the benefit of the whole class will be important role for the teacher.

Talking, listening and reasoning

Figure 1 [6]: Science-based news reports often include the researchers' conclusion or claim. Students could be encouraged to identify and respond to this. Throughout the news-based activity, students will be talking, reading, writing and representing their ideas about the topic as they work together to test their understanding. In addition, the news report might be used to encourage a reasoned response from the reader that could be shared. Most news outlets provide a forum for readers in which to share their views. Students could be encouraged to have their say, make reasoned claims and to comment on the range of responses from other readers. Teachers could choose for this to be either a public or an in-class forum, depending on the students involved. If a public forum is selected, it is important to ensure that appropriate safeguarding is in place.

Multimodal communication

Figure 1 [7]: News text could be used as a stimulus to search for additional information by accessing the full news report and looking for links to other related studies. Students could be encouraged to look for the evidence to support the claim. What evidence would they need? What questions would they ask the researchers? Students could also be challenged to design a sound-related study to test their ideas on noise pollution or sound insulation. Follow-up multimodal writing tasks could illustrate the nature of science communication for students – How could students present the results of their study? What audience would their report target? How could images or graphics support the account?

An important characteristic of a critical reader is to know which news reports are worth paying attention to and which do not stand up to scrutiny. Some science news reports are not useful and sometimes brief and incomplete. It is important to remember that these are features of the report, not the science research. One activity described by Jarman and McClune (2011, p.135) in their guide to using science in the news uses key questions as a framework for analysis of the article as a whole. An exemplar list of key questions is suggested in Figure 2. These can be used to examine the structure, credibility and authority of the report.

In this question framework, evaluation of the structure of the report was based on judgement about its constituent parts, principally the context of the reporting and substance of the report (Q1 and Q2). Its credibility was considered in relation to use of language and the sources upon which it relied (Q3 and Q4). The authority as a reliable source of science communication was judged on the basis of the information provided about the research approach, and the links made between the evidence presented and the reported conclusions (Q5 and Q6).

Literacy in a socio-scientific context

News reports that address socio-scientific issues often place science in a cross-curricular context. *How your clothes are poisoning the world's food supply* (*The Guardian*, 2016) is an attention-grabbing headline (see Figure 3 on p. 220). The science story highlights the problem of plastic waste accumulating in the world's oceans. Images of turtles and sea birds entangled in plastic waste visually convey an emotive message. The news story makes the link between environmental pollution and the food chain. The issue and underlying
question relates to the exponential growth in the use of synthetic polymers in the past 50 years. This has changed the way in which we live, but what can society do about the unanticipated consequences of scientific and technological discoveries?

Figure 2: Evaluating science news reports – key questions

1. **Is there background information about the news report?**

   Think about   Who wrote the newspaper article?

   Does the writer have expertise or special science understanding?

   What media site or newspaper does it appear in?

   Does the media source have a well-known position on the issue?

2. Is the study important?

   Think about   What are the outcomes from this science study?

   How important is it to me?

   How important is it to others in my community?

3. Does the news report appear to be balanced?

   Think about   What words and phrases are being used to influence me?

   What is the balance of fact and opinion?

   Are there alternative points of view?

   Is the language emotive or persuasive?

4. Are the science sources reliable?

   Think about   Who did the research?

   Where was the research done?

   Who funded the research?

   Did the researchers have relevant expertise and experience?

5. How was the research done and reported?

   Think about   Is an experiment described? How was the research carried out?

   Where did the scientists report the results?

   What do other scientists say?

6. Does the evidence support the conclusion?

   Think about   What data were collected?

   Is the evidence given?

   What conclusions were drawn?

   Is there an explanation of the science behind the story?

   How certain are other scientists about the conclusion?

Interestingly, the story has an unexpected twist that illustrates the provisional nature of science and the challenges for a society that depends on it. It would appear that the well-intentioned use of recycled plastic, which improved some manufacturers' eco-friendly standing, could potentially be making the problem worse. How is the reader to respond to this unexpected disclosure?

Figure 3: 'Your clothes are poisoning the world's food supply'

The Guardian (Adapted)   20 June 2016
How your clothes are poisoning the world's food supply

New studies indicate that the fibers in our clothes could be poisoning our waterways and food chain on a massive scale. Microfibers – tiny threads shed from fabric – have been found in abundance on shorelines where wastewater is released.

In an alarming study released Monday, researchers at the University of California at Santa Barbara found that, on average, synthetic fleece jackets release 1.7 grams of microfibers each wash. "These microfibers then travel to your local wastewater treatment plant, where up to 40% of them enter rivers, lakes and oceans," according to findings published on the researchers' website.

Synthetic microfibers are particularly dangerous because they have the potential to poison the food chain. The fibers' size also allows them to be readily consumed by fish and other wildlife. These plastic fibers have the potential to bioaccumulate, concentrating toxins in the bodies of larger animals, higher up the food chain.

In a groundbreaking 2011 paper, Mark Browne, now a senior research associate at the University of New South Wales, Australia, found that microfibers made up 85% of human-made debris on shorelines around the world.

While Patagonia and other outdoor companies, like Polartec, use recycled plastic bottles as a way to conserve and reduce waste, this latest research indicates that the plastic might ultimately end up in the oceans anyway – and in a form that's even more likely to cause problems.

Breaking a plastic bottle into millions of fibrous bits of plastic might prove to be worse than doing nothing at all.

Socio-scientific issues often evoke conflicting responses within a group and this can set up an interesting dynamic in the lesson. Irrespective of the reader's viewpoint, his/her capacity for a reasoned response is an indicator of his/her literacy capability. For the teachers to help their students construct their own reasoned responses involves students using the language of science to enable learning in science. The mastery of the idea is linked to the mastery of the language. Socio-scientific issues often catch students' attention. These are opportunities for students to think and talk about science. A critical response requires the reader firstly to access information, then evaluate the report, before making a reasoned response to the argument being made.

The issue of 'plastic waste' or any similar socio-scientific concern, when integrated with existing curricular topics, links science in school to issues that students care about and respond to with emotion and passion. Unlike most science topics, when exploring these issues students can weigh the evidence, wrestle with the complexity and adopt a personal standpoint. In an attempt to make science personally meaningful, students consolidate and deepen their understanding of the issue. Language is being used to shape both knowledge and understanding of science. Socio-scientific issues are opportunities for the science teacher to expose students to language-rich creative activities, including role play, drama and discussion.

Role play activities, in which students examine the views, attitudes and feelings of participants, are opportunities to promote literacy skills. The plastic waste issue could be addressed, for example, through students taking roles in an investigative television news report that involve journalists, scientists, manufacturers and environmental campaigners. Alternatively, students could plan for their participation in a public meeting as they represent the views of different stakeholders. Participants in the role play will need to be aware of striking a balance between fact and opinion and how to use emotive and persuasive language to support an argument. In preparation for the activity, students could search and analyse the text for different uses of language.

In their guide to science news media, Jarman and McClune (2007) suggest a range of language-based tasks that can shape students' understanding and develop literacy skills as they engage with newsworthy science. Using news-based resources may help students to appreciate how a striking headline, a relevant context, a pertinent analogy or metaphor, or an evocative image, can help the writer to communicate an idea in a memorable way. Other creative writing tasks could involve producing a drama, a science-based story or a case study. Physically enacting a newsroom activity of editing a text to fit a strict word limit with a tight deadline can be an effective way to show students how a science story may be shaped by the constraints of the news outlet and how much of a story gets left on the editor's cutting room floor.

Science communication activities often rely on effective use of images and graphics. Students could examine how images and captions are used in media presentation of science and how these enhance or shape the science narrative. In addition to editing a report, students could select appropriate images and captions to accompany a news report such as those included above. In our news report (Figure 3), the text provides enough information to allow students to create an illustrative graphic to accompany the story, such as a timeline, flow chart or infographic, showing the progression of microfibres from the fleece to the food chain. This could include appropriate labelling and a caption. In translating information from written text to images, students are shaping and consolidating their understanding of science ideas and developing representational literacy skills that are integral to science communication.

In conclusion

Science knowledge and understanding cannot be separated from language that is used to communicate it. Scientists use all the modes of communication highlighted in this chapter, so all of them should have a place in learning and doing science in school. Students, at every stage in their development, will be able to extend their capacity in literacy and need a literacy challenge and an opportunity to learn about and develop their capacity for multimodal science communication. The ability to engage with media-reported science through a critical scientific literate eye, and to make a personal and reasoned response, could be an enduring outcome of an education in science and one that should be valued.

Further reading

Jarman, R. & McClune, B. (2007) *Developing Scientific Literacy*. Maidenhead: Open University Press

McClune, B. & Alexander, J. (2015) 'Learning to read science-related media reports with a critical eye: cultivating discerning readers of media reports with a science component', *School Science Review*, **97,** (359), 15–20

McClune, B. (2017) 'Making the most of the News: approaches to using media based learning contexts'. In *Contextualizing Teaching to Improving Learning: The case of Science and Geography*, Leite, L., Dourado, L., Afonso, A. & Morgado, S. (Eds.). New York: Nova Science Publishers

Wellington, J. & Osborne, J.F. (2001) *Language and Literacy in Science Education*. Buckingham: Open University Press

Yore, L. (2012) 'Science literacy for all – more than a slogan, logo or rally flag!'. In *Issues and Challenges in Science Education Research: Moving Forward*, Tan, K.C.D., Kim, M. & Hwang, S. (Eds.). Dordrecht: Springer

References

Council for Curriculum, Examinations and Assessment (CCEA) (2012) *GCSE Science Single Award Specification*. Available at: http://www.rewardinglearning.org.uk/microsites/general_science/single_award/specification/index.asp Accessed 15.06.17

Department of Education (2014) *Science Programme of Study at Key Stage 4*. Available at: https://www.gov.uk/government/uploads/system/uploads/attachment_data/file/381380/Science_KS4_PoS_7_November_2014.pdf Accessed 15.06.17

Education Scotland (2010) *Curriculum for Excellence: Science.* Available at: https://education.gov.scot/Documents/sciences-pp.pdf Accessed 15.06.17

Evagorou, M. & Osborne, J. (2010) 'The role of language in the learning and teaching of science'. In *Good Practice in Science Teaching: What research has to say,* Osborne, J. & Dillon, J. (Eds.). Maidenhead: Open University Press

Hand, B. & Prain, V. (2002) 'Teachers implementing writing-to-learn strategies in junior secondary science: A case study', *Science Education,* **86**, (6) 737–755

Harlen, W. (2010) *Principles and Big Ideas in Science Education.* Hatfield: Association for Science Education. Available at: https://www.ase.org.uk/resources/big-ideas/ accessed 05/05/2011

Hines, P.J., Wible, B. & McCartney, M. (2010) 'Learning to read, Reading to learn', *Science,* **328,** (5977), 447

Jarman, R. & McClune, B. (2007) *Developing Scientific Literacy.* Maidenhead: Open University Press

Jarman, R. & McClune, B. (2011) *Science Newswise 2,* Hatfield: Association for Science Education

Jenkins, E. (2010) 'How might research inform scientific literacy in schools?', *Education in Science,* (239), 26–27

McClune, B. (2017) 'Making the most of the News: approaches to using media based learning contexts'. In *Contextualizing Teaching to Improving Learning: The case of Science and Geography,* Leite, L., Dourado, L., Afonso, A. & Morgado, S. (Eds.). New York: Nova Science Publishers

McClune, B. & Jarman, R. (2012) 'Encouraging and equipping students to engage critically with science in the news: What can we learn from the literature?', *Studies in Science Education,* **48,** (1), 1–49

National Research Council (2012) *A Framework for K-12 Science Education: Practices, Crosscutting Concepts, and Core Ideas.* Committee on A Conceptual Framework for New K-12 Science Education Standards. Board on Science Education, Division of Behavioural and Social Sciences and Education. Washington, D.C: The National Academies Press

NGSS (2013) *New Generation Science Standards.* Washington DC: The National Academies Press

Norris, S. & Phillips, L. (2003) 'How literacy in its fundamental sense is central to scientific literacy', *Science Education,* (87), 224–240

OECD (2015) *PISA 2015 Assessment and Analytical Framework Science, Reading, Mathematic and Financial Literacy.* Available at: http://www.oecd-ilibrary.org/docserver/download/9816021e.pdf?expires=1497555080&id=id&accname=guest&checksum=98B92FB84427803D7B20ABB74B800C16 Accessed 15.06.17

Pearson, P.D., Moje, E. & Greenleaf, C. (2010) 'Literacy and Science: Each in the service of the other', *Science*, (382), 459–462

Prain, V. & Tytler, R. (2012) 'Learning science in school through constructing representations'. In *Spectra: Images and data in Art/Science*, Kennedy, C. & Rosengren, M. (Eds.), Proceedings of the SPECTRA 2012 symposium, Canberra, Australia

Science Daily (2016) *Traffic noise reduces wild owls' foraging ability.* Available at: https://www.sciencedaily.com/releases/2016/09/160927082744.htm Accessed 15.06.17

The Guardian (2016) *How your clothes are poisoning the world's food supply.* Available at: https://www.theguardian.com/environment/ 2016/jun/20/microfibers-plastic-pollution-oceans-patagonia-synthetic-clothes-microbeads Accessed 15.06.17

Their, M. (2002) *The New Science Literacy.* Portsmouth, NH: Heinemann

Waldrip, B., Prain, V. & Carolan, J. (2006) 'Learning junior secondary science through multimodal representations', *Electronic Journal of Science Education*, **11**, (1), 87–107

Wellington, J. & Osborne, J.F. (2001) *Language and Literacy in Science Education.* Buckingham: Open University Press

Yore, L. (2012) 'Science literacy for all – more than a slogan, logo or rally flag!'. In *Issues and Challenges in Science Education Research: Moving Forward*, Tan K.C.D., Kim, M. & Hwang, S. (Eds.). Dordrecht, Netherlands: Springer

SECTION 4:

Assessing science

CHAPTER 18

A potted history of AfL in science

Andrea Mapplebeck & Chris Harrison

This chapter provides an overview of assessment for learning (AfL), also called formative assessment, over the last twenty years, exploring the purposes of assessment and how different strategies can be used by teachers to evaluate students' learning and hence inform subsequent teaching, and discussing the role that students can play in assessing their own understanding.

Introduction

In recent years, the Ofsted inspection handbook has broadened out its criteria for evaluating the quality of teaching within a school from 'teaching' in 2014 to 'teaching, learning and assessment' in 2016, highlighting the increased importance and interconnectivity in the nature of teaching, learning and assessment. With teachers working harder than ever before, there is a need to ensure that their use of formative practices are relevant and beneficial, and that they optimise the time spent in promoting learning and outcomes for students.

It is nearly two decades since Black and Wiliam (1998) published *Inside the Black Box* and introduced a generation of educators to 'AfL' (Assessment for Learning), and the ideas and concepts that it encompasses. However, Black (2010) reported that AfL was not happening in a large number of classrooms, due to the failure of many teachers to properly grasp the concept and the underlying principles and strategies it advances. Black (2010) also expressed a concern that there are still many teachers who are confused by the different purposes of assessment, how to use it for their respective purposes and, in the case of assessments that function formatively, how to respond to the evidence elicited.

With education and schools evolving considerably since the publication of Black and Wiliam's (1998) initial review, the question remains whether AfL still has a part to play in improving the practice of teachers and the subsequent outcomes of students. Black *et al* (2002) argued that enhanced formative assessment practices could produce gains in student achievement, even when measured in such narrow terms as National Curriculum tests and examinations. In order to attain these increases in educational achievement, Wiliam (2011) highlighted the need to support teacher professional development that focuses on how to develop minute-by-minute (i.e. within lesson) and day-to-day (i.e. between lessons) formative assessment practices. There remains a need to raise teacher awareness of what is involved when AfL practices are incorporated into teaching (Harrison, 2017) and also a greater focus on the important role that students play in this formative process.

Purposes of assessment

Teaching and learning are interdependent activities and, for the teacher and the student to be aware of progress, and what the implications are for the next steps in learning, some form of assessment needs to be undertaken. The etymology of the word 'assessment' derives from the Latin 'assidere', meaning to 'sit by' – an indication of the synergy between the teacher and the student during teaching, learning and assessment. Gibbs (1999) contends that 'assessment is the most powerful lever teachers have to influence the way students respond to the course and behave as learners' (p. 41). As such, assessment is an integral part of all aspects of school life (Tarras, 2005) and therefore a fundamental aspect of a teacher's repertoire, being used daily both consciously and subconsciously as they interact with their students.

Assessments can be formal or informal, take place during or at the end of a sequence of learning activities, and be presented in either written or oral form, and so there are many choices for the teacher to make. To elicit evidence of a learner's understanding, development of skills or accumulation of knowledge, there is a wide range of different approaches that teachers can utilise as part of their assessment repertoire, including: diagnostic activities; probing and recall questions; teacher-generated tests; externally-generated tests; and authentic assessment activities. However, it is what is done as a consequence of the assessment that is of utmost importance and determines whether or not it functions summatively or formatively.

Hattie (2003) claims that the primary concern of assessment is to provide feedback to teachers and students. Torrance and Pryor (2001) highlight issues that exist with teachers' everyday use of assessment. They suggest that teachers can employ assessment continuously within their learning environment, which leads them to perceive that they are using it formatively. Nonetheless, there is the possibility that these assessments being used in a sustained manner are actually summative in nature; that is, the teacher elicits evidence without actually either responding to it, or using it to impact on their practice or student thinking. Therefore, it cannot be claimed that, just because a teacher uses strategies such as mini-whiteboards, traffic lights, exit tickets, wait time, no-hands up, etc., they are employing formative assessment in their classroom.

Assessment *for* learning focuses on where students are in their learning and how they might be helped in moving that learning forward. In AfL, teachers use a variety of questions, tools and practices to find where students are in their learning. So, in the topic of forces, a question such as 'What unit do we measure force in?' is a simple recall-of-knowledge question, whereas 'What do you think friction is like on the Moon?' is a far more probing question, which is likely to inform teachers of current understanding and also any alternative conceptions that students may have (Black & Harrison, 2004) about challenging ideas such as friction and gravity. Therefore, the latter question has far more formative potential than the recall question.

Some assessment tools can be used throughout a sequence of learning to provide an ongoing idea of developing understanding. For example, teachers might use KWL grids. These are worksheets with 3 column tables that ask the learners to list first what they feel they **Know** at the start of the activity, what questions they might have (**Want to know**) and, finally, at the end of the activity, what they think they have **Learned**. These types of activities encourage learners to talk and express their understanding as they undertake classroom activities, providing the teacher with assessment evidence at the start and during those activities. The successful AfL teacher is *'able to ask the right questions at the right time, anticipate conceptual pitfalls, and have at the ready a repertoire of tasks that will help students take the next steps. This requires a deep knowledge of subject matter'* (Shephard, 2000, p.12).

For teachers, the substantive potential of these AfL practices to inform student learning actions depends on what teachers notice and select as a focus from the assessment evidence (Mason, 2002; Ball, 2011), and how they interpret and act on the information they have (Cowie & Harrison, 2016). AfL therefore could be described as responsive teaching, the subsequent actions coming from either the teacher or the student as a consequence of what has been inferred from the assessment evidence arising through talk and activities.

Skidmore (2000) also showed that there are times in the classroom discourse when teachers have alternative choices, requiring them to face what he calls 'critical turning points', when they have the opportunity to push the interaction in one direction or another. Sometimes teachers lead students back to the intended learning goals, while at other times they might focus on and develop further ideas raised by students. The following vignette, from a Year 8 (age 13) classroom, illustrates the choices and actions that one teacher made.

She had been introducing 'working scientifically' into her lessons and was attempting an activity that would introduce her class to the 'messiness of data'. The question she set her class was: 'How many drops of water can you get on a two pence piece?' Students worked in groups of three and the following transcript was collected early in the lesson, as groups used pipettes to drop water onto a coin:

Student 1: Miss, we've got twenty.

Teacher: Twenty drops. That's great. Do you think it will take more? You said more, Cassandra. Why did you think that?

Students: Oww! *Water runs off the coin.* 23, Miss! It took 23.

Student 2: Got ridges and sticky-up bits that will keep the water in place.

Teacher: So, would it hold the same amount of water on the other side?

Student 1: The tails? Shall we try?

Teacher: What are you going to try and why?

Student 3: The other side to see if it has the same number [of drops].

Student 2: Yes, the pattern is different and so that will affect it.

Here we see that the teacher begins the interrogation of student ideas by asking questions about the data they have collected. However, Student 2 starts to focus on a variable – whether the tail or head of the coin is being used. The teacher notices this and recognises it as an opportunity to pursue the group's thinking about manipulating variables. Later in this lesson, the teacher returns to her original objective and focuses the whole class on the variability of their data and the importance of repeats in a practical activity. She then brings in the idea of control and manipulation of variables that the group of students in the above transcript introduced. In attending to what arises in the classroom as well as the learning objectives for the activity, responsive teaching looks for connections between (1) the students' reasoning and actions and (2) core ideas and practices of the discipline, (Robertson *et al*, 2015). This approach forges a way forward that helps students to engage with science ideas and practices so that they can hone and develop their understanding. The challenge for teachers is in how to manage the interaction between student experience and disciplinary ideas (Thompson *et al*, 2016).

The role of the student in AfL

Classrooms have become more interactive learning environments, where it is acknowledged that students are not empty vessels waiting to be filled with knowledge and where identifying current thinking is needed before new thinking is introduced. This is especially important for science teachers, as students are regularly interpreting the world around them, which often means that they start from alternative conceptions for the science phenomena they encounter; how heavy and light objects fall, whether plants respire, or that all metals are magnetic are prime examples! Without taking time to elicit where the students are with their thinking, valuable opportunities to challenge and move forward scientific understanding can be missed, which may impede future progress.

Many teachers try to engage students more actively in their learning by sharing with their students the purpose of the lesson and criteria for success. However, in such environments, the responsibility for the use of assessments often remains very much with the teacher. In some instances, this can lead to passive and teacher-reliant students who need constant direction and assurance. In classrooms where AfL is embedded, there is a shift in the culture towards what Blanchard (2009) classes as transparency. Transparency involves increasing students' autonomy so that they are active and responsible co-owners, negotiating with the teacher various aspects associated with their learning. One way to conceptualise this is to think about how we learnt to drive – this was not by sitting in the passenger seat. This transparency occurs when there is dialogue between the teacher and the student linked to learning and not tasks, and where assessment and feedback are used to help improve performance and recognise progress during the learning.

In order to ensure that assessment becomes a more collaborative and interactive enterprise, teachers need to provide opportunities with their students to discuss, clarify and negotiate the learning, drawing on these various approaches. Five key classroom strategies have been identified as being associated with this process. These are:

1. Clarifying and sharing learning intentions and criteria for success.
2. Engineering effective classroom discussions and other learning tasks that elicit evidence of student understanding.
3. Providing feedback that moves learners forward.
4. Activating students as instructional resources for one another.
5. Activating students as the owners of their own learning.

Table 1 opposite shows how classrooms can move from being passive to dynamic learning environments for students. These ideas are linked to each of the five key strategies, with further examples of how this can be achieved in science lessons.

An additional benefit of creating more dynamic classrooms is that students see that their ideas are sought and responded to, and that classroom interactions focus less on being correct or wrong and more on what is or is not understood. Students see the teacher valuing their input and adapting the teaching to help

Table 1: How classrooms move from passive to dynamic learning environments

Key Strategy	Passive Classrooms	Dynamic Classrooms	Example of Dynamic Activity
Clarifying and sharing learning intentions and criteria for success	The teacher shares with the students the learning goals and success criteria for an activity.	Students are given a range of different exemplars and have to decide together the criteria for a quality piece of work.	Provide a number of different completed answers for the same question and the mark allocation. Students award marks to the completed answers and give reasons for their choices. They attempt several different questions in the same context.
Engineering effective classroom discussions and other learning tasks that elicit evidence of student understanding	The teacher asks closed, recall questions to students who put their hands up.	Students are presented with an event (via video or demonstration) related to the topic and their prior learning and asked to come up with their own explanation.	Students watch the egg in the conical flask demonstration (https://youtu.be/uzJ0CqUD12I) and work in groups to develop an explanation using the particle model of matter.
Providing feedback that moves learners forward	Teachers tell students what they need to do to improve their work.	The teacher asks questions or provides feedback that focus the student on an incorrect, ambiguous or poorly explained idea.	An error with scaling on a graph axis is noticed by the teacher. The teacher says: 'Have a look at your graph and the success criteria. There's an error. Try and work out what it is and correct it.'
Activating students as instructional resources for one another	The teacher shares with the students the learning goal and success criteria for a practical activity and models the process to the students.	In small groups, students are provided with a range of equipment and discuss which they think is the best approach to undertake to achieve the learning goal for a practical activity.	Before doing a rates of reaction investigation, students explore and compare different ways of data collection. They decide and give reasons for which approach they think will produce the most accurate and reliable results. They share their reasoned decisions with other groups.
Activating students as the owners of their own learning	Teachers assess students and direct them as to where they should start the lesson.	All students work on the same learning intention and choose from an increasing level of challenge where they wish to start the activity.	Students are learning to balance chemistry equations and shown a list of 15 different examples that increase in difficulty. They are asked to complete 6 of their own choice.

all students to move forward in their learning, and consequently feel more motivated. This approach builds student agency and, as such, contributes to lifelong learning skills.

In conclusion

AfL practices have had a massive impact on the way in which teachers plan, organise and reflect, making science activities more interactive, and this move has supported and engaged students more in science learning. While our formal examinations often focus more on recall, comprehension and application than on the higher thinking skills of analysis, evaluation and creating, working in a more formative manner in classrooms enables teachers to prepare students well for examinations, while at the same time instilling many of the skills, attributes and practices of scientists. However, perhaps one of the key motivators for teachers to adopt and adapt a formative approach is that they learn much more about how and what their students are learning, and so assessment for learning can lead to teacher professional learning as well.

Further reading

Harrison, C. (2015) 'Assessment for Learning in STEM Classrooms', *Journal of Research in STEM Education,* **1,** (2), 78–86. Available at: j-stem.net/wp-content/uploads/2016/12/Harrison_C_01.pdf

References

Ball, D. (2011) 'Foreword'. In *Mathematics Teacher Noticing: Seeing through teachers' eyes,* Sherin, M., Jacobs, V. & Phillip, R. (Eds.), pps. xx-xxiv. New York: Routledge

Black, P. & Wiliam, D. (1998). *Inside the Black Box: Raising standards through classroom assessment.* London: School of Education, King's College London

Black, P. (2010) 'A decade on, academic who led AfL movement admits it "isn't happening" in many classrooms', *TES*

Black, P. & Harrison, C. (2004) *Science inside the Black Box.* London: nferNelson

Black, P., Harrison, C., Lee, C., Marshall, B. & William, D. (2003) *Assessment for Learning – putting it into practice.* London: Open University Press

Black, P., Harrison, C., Lee, C., Marshall, B. & Wiliam, D. (2002) *Working Inside the Black Box.* London: King's College London Department of Education and Professional Studies

Blanchard, J. (2009) *Teaching, Learning and Assessment.* Maidenhead: Open University Press

Cowie, B. & Harrison, C. (2016) 'Classroom processes that support effective assessment'. In *Handbook of Human and Social Conditions in Assessment,* Brown, G.T.L. & Harris, L.R. (Eds.), pps. 335–350. New York: Routledge

Gibbs, G. (1999) 'Using assessment strategically to change the way students learn'. In *Assessment Matters In Higher Education,* Brown, S. & Glasner, A. (Eds.). Buckingham: Society for Research into Higher Education and Open University Press

Harrison, C. (2015) 'Assessment for Learning in STEM classrooms', *Journal of Research in STEM Education,* **1,** (2), 78–86 (online)

Harrison, C. (2017) 'Adapting pedagogy for formative assessment'. In *Encyclopedia of Educational Philosophy and Theory,* Peters, M. (Ed.). USA: Springer

Hattie, J. (2003) *Formative and summative interpretations of assessment information.* Auckland, New Zealand. Available at: https://cdn.auckland.ac.nz/assets/education/hattie/docs/formative-and-summative-assessment-(2003).pdf

Mason, J. (2002) *Researching your own practice: The discipline of noticing.* Abingdon: Routledge

Mason, J. & Davis, B. (2013) 'The importance of teachers' mathematical awareness for in-the-moment pedagogy', *Canadian Journal of Science, Mathematics and Technology Education,* **13,** (2), 182–197

Robertson, A., Richards, J., Elby, A. & Walkoe, J. (2015) 'Documenting Variability Within Teacher Attention and Responsiveness to the Substance of Student Thinking'. In *Responsive Teaching in Science and Mathematics,* Robertson, A., Scherr, R. & Hammer, D. (Eds.), pps. 227–248. Abingdon: Routledge

Sheppard, L. (2000) 'The role of assessment in a learning culture', *Educational Researcher,* **29,** (7), 4–14

Skidmore, D. (2000) 'From pedagogical dialogue to dialogic pedagogy', *Language & Education,* **14,** (4), 283–296

Tarras, M. (2005) 'Assessment – summative and formative – some theoretical reflections', *British Journal of Educational Studies,* **53,** (4), 466–478

Thompson, J., Hagenah, S., Kang, H., Stroupe, D., Braaten, M., Colley, C. & Windschitl, M. (2016) 'Rigor and responsiveness in classroom activity', *Teachers College Record,* **118,** (5)

Torrance, H. & Pryor, J. (2001) 'Developing formative assessment in the classroom: Using action research to explore and modify theory', *British Educational Research Journal*, **27,** (5), 615–631

Wiliam, D. (2011) *Embedding formative assessment.* Bloomington: Solution Tree Press

CHAPTER 19

Using data to inform science teaching

Pete Robinson

This chapter is intended to encourage teachers to think about the different types of data that might be available, how teachers and others might use the data and to consider the limitations of data. We look at the range of numerical data available for teachers to use, the ways in which they are held accountable through these data, and productive ways of using these data to improve teaching and learning.

Introduction

Teachers can access large amounts of data about their school, other schools in the area and their students' backgrounds and attainment. Some of these data are also generated by teachers and reported, whether to the students themselves, parents, Senior Leadership Teams (SLTs), or published more widely. It is therefore important for teachers to think about what data they are collecting and how they are collecting them, what data may tell them and how to make sure that data they collect are useful and relevant.

The Assessment Pyramid

A useful way of thinking about assessment and data is by considering an Assessment Pyramid.

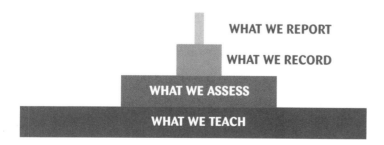

'What we teach' refers to all activities in the classroom during which students are learning new skills and knowledge, or are using existing skills and knowledge to consolidate and extend learning.

'What we assess' is mainly assessment in the classroom based on students' responses, and feedback that we use formatively during lessons (see Chapter 18). With effective assessment for learning, what we assess may be extensive but fleeting, except for written feedback. 'What we assess' also includes more formal assessment of tasks, homework, tests and examinations. Assessing and recording assessment outcomes can be a time-consuming process and so it is important that we assess and record useful information.

'What we record' may include outcomes from individual or collaborative tasks, homework, tests and examinations. It may include information about students' efforts, attainment and progress and may be recorded numerically, using letters and comments, and evidenced by photograph, video and artefact. However, the assessment outcomes we choose to record must be recorded in a way that is meaningful to everyone.

'What we report' refers to assessment information that we pass on to the SLT, parents, and other agencies such as academy trusts, local authorities, examination boards and the Department for Education in the context of England.

Quite clearly, what we record and report is a tiny proportion of what we teach and assess. This is because it is unnecessary to record and report every minor achievement made by students: what we record and report is a 'barometer' of achievement. It must therefore be useful for a wide range of stakeholders and representative of the students' abilities.

A recent history of assessment in science

Prior to the National Curriculum being introduced in 1989 to England, Wales and Northern Ireland, schools were free to construct the curriculum that best suited the needs of their students. For example, state grammar schools selected the 'most able' students and provided a largely academic curriculum, mainly leading to GCE 'O-level' (General Certificate of Education Ordinary Level) qualifications. The definition of 'most able' in this context is *academic students most capable of passing the selection examination*' at ages 11 or 13. In contrast, secondary modern schools often provided a more vocational education, mainly leading to CSE (Certificate of Secondary Education). During the 1970s, most local authorities converted selective schools into comprehensive schools and the GCSE (General Certificate of Secondary Education) for all students was introduced for first examination in 1988.

Most schools provided a coherent and suitable curriculum, which was shaped by the qualifications offered to the students and often by the school context.

While examination boards provided a syllabus for each subject that identified what should be taught for most students, schools were allowed to develop their own curriculum and assessments, which were then vetted and moderated by one of the 29 examination boards. This meant that schools in agricultural areas, such as Norfolk, could design a curriculum that suited the life experiences and aspirations of their students; similarly, a school in Tower Hamlets, London, could devise a different curriculum for their students because of the very different lifestyles of those students. However, this was problematic for students who moved during their school years and could be difficult for school leavers who didn't 'fit' their local mould. Past papers provided insight into skills that were needed, and a department would incorporate relevant skill teaching in the curriculum from Year 7 (age 12). Terms such as 'assessment for learning' and 'intervention' were not used, but the processes were.

Recording student outcomes was largely done by letter grade or mark, which offered little insight into students' knowledge and skills. Departments with competent and experienced staff functioned well with this system and inducted new and inexperienced members of staff. There were, however, systemic weaknesses, which the National Curriculum was intended to address.

Levels

National Curriculum levels were first introduced in 1988 along with the first National Curriculum. There were initially ten levels of assessment across all four Key Stages, based on ideas worked out by the Task Group on Assessment and Testing (see www.educationengland.org.uk/documents/pdfs/1988-TGAT-report.pdf), but these were later reduced in number to eight levels for Key Stages 1-3 (ages 5-14), with an additional level for *exceptional performance*. Levelness across different subjects was achieved, as all National Curriculum subjects had level descriptions based loosely upon Bloom's Taxonomy. The National Strategies, introduced in 1997, created a national CPD programme for all schools, which aimed to increase teachers' skill levels in recognising the level at which students were working based upon general level descriptions. Shortly prior to the end of the National Strategies in 2011, Assessing Pupils' Progress (APP) was introduced, which increased the complexity of assessment and recording but provided schools with a powerful assessment tool and a rich source of data. In the schools that embraced APP, learning became much more open-ended, student-centred and skill-based, and there was enthusiasm for the potential that this had to transform the curriculum (Stutchbury, 2011). However, relatively few schools welcomed APP due to procedural and cultural concerns (Hillier, 2012): these included the time required to assess and record students' work, at an individual level, across 109 assessment criteria across six levels; the extent to which tasks were being 'bolted on' rather than embedded; the lack of moderation; and the extent to which schools could take ownership of the process. It became clear that there had been *'insufficient understanding of the complex and lengthy process required for schools to change from a culture of high-stakes external assessment to low-stakes internal assessment'* (Hillier, 2012 p.33).

Assessment without Levels (AwL)

The 2014 National Curriculum was published without level descriptors, and levels were no longer used for statutory assessments from September 2015. There was a clear rationale for this decision, as explained by the Commission on Assessment without Levels' *Final Report* (2015), led by John McIntosh CBE: *'Despite being intended only for use in statutory national assessments, too frequently levels also came to be used for in-school assessment between key stages in order to monitor whether pupils were on track to achieve expected levels at the end of key stages. This distorted the purpose of in-school assessment, particularly day-to-day formative assessment. The Commission believes that this has had a profoundly negative impact on teaching.*

'Too often, levels became viewed as thresholds and teaching became focused on getting pupils across the next threshold instead of ensuring they were secure in the knowledge and understanding defined in the programmes of study. Depth and breadth of understanding were sometimes sacrificed in favour of pace. Levels also used a "best fit" model, which meant that a pupil could have serious gaps in their knowledge and understanding, but still be placed within the level. This meant it wasn't always clear exactly which areas of the curriculum the child was secure in and where the gaps were' (p.5).

The legacy from the decision to abandon levels is a wide range of practice regarding assessment and recording of data. For example, some schools continue to use and share levels with students, whilst others use levels and sub-levels internally for tracking and without sharing them with students or parents. Many schools use broad descriptions such as such as *'making expected progress'*, *'making less than expected progress'* and *'exceeding expected progress'*. Assessment without Levels has paved the way for identifying long-, medium- and short-term curricular (learning) targets for students, as well as recognising and celebrating skills and knowledge that they have secured.

Most schools test students on a regular basis, recording marks in a variety of different ways, including raw scores, percentage scores, estimated grades or levels (despite 'levels' having been abandoned). Some schools spend 10% or more of curriculum time testing or preparing students for tests (Ofsted, 2013). Often, test and examination marks are recorded and only used to gauge whether students are making expected progress. Using tests formatively is excellent practice and is used well by some science departments. For example, short- and medium-term curricular (learning) targets may be determined from script analysis or mark analysis of topic tests, and longer-term targets from end-of-year examinations. The challenge for departments is how to communicate these in an effective manner, and how to address them.

In a similar way to recording levels at Key Stage 3 (age 11-14), GCSE grades may be recorded at Key Stage 4 (age 14-16). Grading Key Stage 4 tests and examinations is more difficult than levelling Key Stage 3 tests and examinations, because GCSE grades are cohort-referenced and not criterion-referenced: grade boundaries are determined statistically by the examination board and grades determined after all examination scripts have been marked and standardised. However, skilled and experienced secondary science teachers can judge grade thresholds of examinations and tests fairly accurately, based on the performance of key students. Changes to GCSE grading from A*-G to 9-1 is currently making predictions of grade thresholds more difficult, although the situation is likely to improve with time.

It is important to emphasise that marks, grades or levels obtained from tests and examinations do not identify curricular (learning) targets or inform intervention needs: these require script analysis and/or discussion with students. For example, if most students score low marks on a particular question, the reason *why* they scored low marks might be mathematics-related, a literacy issue or caused by weak science understanding, and this might only be revealed through discussion with individual students. Well-constructed test papers should contain a good balance of recall, application, and analysis and synthesis questions to match terminal assessment. They should also contain a good mix of content, inquiry and maths-based questions on which to base intervention activities.

Accountability measures

Accountability measures, introduced by recent successive governments to make school performance information available for public scrutiny, have increased schools' focus on collecting and using data in their everyday practice. Success in, and especially by, schools is seen almost exclusively as an increase in examination results. Student skills such as confidence, adaptability, politeness and resilience are difficult to measure and are not explicitly valued as much as they should be. The consequence of high-stakes examinations has led some schools to value results over other indicators of student learning, such as students' enjoyment of subjects and their engagement in learning. In some cases, schools have drastically modified their curricula to focus on what they believe is important in terms of examination success. For example, in science, assessment of practical skills is now done through written examination questions based on a set number of specific practicals. Some schools now focus almost solely on covering the skills that may come up in the examination and this may deter from developing conceptual understanding through engagement with practical work.

Accountability measures themselves deserve some discussion, given that they rely heavily on certain examination results. This sometimes means, indirectly, that changes to these measures can have a rapid and significant impact on the curriculum, on teachers and on students within many schools. For example, the change from 'Five or more A*-C grades including English and Maths' to 'Progress 8' forced schools to radically rethink which students to target for intervention. Previously, C/D borderline students were targeted *irrespective of their starting point*. 'Progress 8' means that under-performing students of all abilities are now targeted. The 'Progress 8' measure requires subjects to be collected in 'buckets'. 'Bucket Two' can contain science GCSE subjects, but not science BTEC, which can only count in 'Bucket Three'. One consequence is that many schools that traditionally offered BTEC science no longer do so. This might restrict curriculum choice offered to students, with science opportunities being ruled by demands from other areas of the curriculum. In 2013, Ofsted reported that:

'Although different courses with different assessment arrangements may appear to meet a range of learning needs, in practice it is not the course specification but the effectiveness of the teaching that engages students. It is possible for a predominantly academic Key Stage 4 pathway to meet all students' learning needs in science and enable them to progress to all 16+ science pathways, including employment or apprenticeships. In the best lessons seen, the teachers made the course content come alive and pushed learning well beyond the specifications' (Ofsted, 2013 p.36).

If forced to provide a narrow science curriculum, it is essential therefore that science departments evaluate the appropriateness of the schemes of learning for all students, and consider modifying them to maintain students' engagement.

Accountability measures have also made an impact on the way in which some schools deliver the curriculum. For example, some schools 'play it safe' by ensuring that everything on the specification is taught to all groups irrespective of whether students understand it. Equally, they do not stray from the specification by looking at science in the news or science related to local contexts, which may be highly engaging for their students. Ironically, some schools aiming for good science results to satisfy accountability measures fail to inspire their students and this has an effect on uptake of science at higher levels of education:

'Getting the grade is not the same as "getting" the science. Too frequently, GCSE grades indicated that students were doing well but they were not enjoying science. Because the subject is a statutory requirement to 16 in local authority maintained schools, and is made compulsory by the academies visited, most students have no choice but to continue studying science at Key Stage 4, and most want to do their best. But once they can choose other courses at 16, most students drop science completely. Despite some 316,000 students nationally achieving two or more good GCSE grades in science in 2010, only 80,000 went on to study one or more advanced level sciences in 2012. This represents a major loss of science talent' (Ofsted, 2013 p.26).

Accountability measures in schools have led to more emphasis on student tracking, predicting external assessment outcomes and targeting students with intervention activities, should their apparent progress be less than that predicted by the school's system. In schools, this system is frequently supported through a commercially available algorithm, such as the Fischer Family Trust (FFT), which predicts GCSE outcomes based predominantly on Key Stage 2 (age 7-11) cohort data results. Student GCSE outcomes are predicted by comparing attainment in a school with a large historical dataset. The advantage of using an algorithmic prediction method is that it draws from a large data set to make predictions. FFT refines the process still further by offering different prediction models for schools to choose from. Factors taken into account may include gender, socio-economic status, SEND (Special Educational Needs and Disabilities), ethnicity and EAL (English as an Additional Language). It will therefore predict school results for a particular cohort. The disadvantage of such methods is that it cannot predict outcomes for individuals, nor can it address deficiencies in teaching and learning.

The journey from Key Stage 2 results to predicted levels or grades throughout school are sometimes referred to as 'flightpaths' against which students are measured throughout their school careers. As Key Stage 2 data have become less detailed, schools may use different methods for identifying the starting point on a flightpath. For example, some schools are using Cognitive Ability Tests (CATs), whereas others are using subject-related pre-testing. Both these methods have strengths and weaknesses. For example, subject-related assessment may depend on the quality of science provision at Key Stage 2 more than students' potential. CAT tests are not subject-specific and so ignore students' strengths and abilities in science. It is particularly unfortunate that both these methods are very stressful for students during transition from primary school to secondary school, a process that may already be causing them anxiety.

Because National Curriculum levels should no longer be used (Harrison, 2016; Commission on Assessment without Levels, 2015), some schools have chosen to use GCSE 'grades' from Year 7 throughout the school to produce flightpaths for students. There are broadly two methods in use within schools. The first uses a method whereby Year 7 students are assigned a target grade for the end of Year 11 (age 16) determined from a baseline assessment. Provided the student makes expected progress, the target remains constant throughout the student's journey through school. The second method grades students according to how they would perform in a hypothetical GCSE at that time, even though they would not have covered the necessary work. In this method, a student's attainment grade would be likely to increase by 2-4 grades throughout his/her school journey. Sharing either method with students is unlikely to motivate them in the same way as positive, specific written feedback towards curricular (learning) targets set for them.

While the focus of assessment data is often on the individual, much of the learning that is done in schools is collaborative. Collaborative tasks are used by some schools as starting point activities to identify students' prior knowledge or 'working scientifically' skill levels in science. Such tasks support learning, as they allow students to draw on one another as a resource and tend to engage students more with scientific ideas. Collaborative tasks are much less stressful for students and provide a rich source of information upon which to base intervention activities throughout the unit of work. Identifying suitable starting points and then building on these throughout the unit of work is recognised as good practice by Ofsted:

'However, despite the attainment data available from frequent topic tests and from earlier stages, too many lessons in the schools visited did not take enough account of what students had already learnt and where they needed further teaching. In a third of the lessons seen, the activities provided for some did not match their learning needs well enough to ensure that they made good progress. In practice, despite the widespread "differentiation by outcome" seen in the survey schools, once students began work, teachers adjusted their teaching quickly. It would make a significant difference to raising expectations if teachers assigned all students to the most challenging tasks from the outset, before adjusting them if they turned out to be too demanding. The general assumption, however, was that "pupils can't", rather than "pupils can". This was seen to be most damaging at the start of Year 7' (Ofsted, 2013 p.31).

Teachers need to know how well their students are learning, where they are in their learning for that topic, and what activities, challenges and support they need next to increase their learning, as, indeed, do the students (see Chapter 18).

Using data from tasks

Classwork and homework tasks are routinely assessed and outcomes recorded as an effort and/or an attainment mark. Attainment may be recorded in different formats according to schools' marking and assessment policies. Some of these data might be useful in judging whether students are making expected progress, but the real value of tasks is that they provide an opportunity for dialogue between students, and between students and their teachers. The dialogue may be through verbal or written feedback, or both. Open-ended classwork and homework tasks provide particularly rich opportunities for identifying learning (curricular) targets. Clearly, such targets could be science knowledge-related. For example, students may understand the principle of the reflex arc without being able to state key vocabulary. Similarly, targets may be skill-related, such as graph skills, application of knowledge or analysis of data. Sometimes, targets might be literacy-based, such as using command words to decode questions, or constructing extended written explanations.

Feedback from online assessment tools

Many schools use online assessment tools to support homework activities, revision and testing. Such tools are increasing in popularity amongst teachers, because they have the potential to provide specific assessment information and are marked automatically and often immediately. Many are also popular with students, as they may be allowed to use iPads or their phones; some students may also like the competitive nature of some of the quiz activities and welcome the immediate feedback. However, most assessment tools assess knowledge rather than understanding and so operate at lower Bloom's Taxonomy levels. It is important to realise that many online assessment tools cannot assess extended writing skills, or higher Bloom's Taxonomy skills such as analysis, synthesis and evaluation. So, what this means is that the assessment tool may give an analysis of a student's knowledge without being able to report on that student's understanding.

In conclusion

Assessment is a time-costly exercise, so it is important that teachers make best use of the data they collect to ensure that assessments are used for formative and summative purposes. Although we are encouraged to assess without levels, it is still essential to consider what progression means in science from Year 7 to Year 11 and ensure that there are opportunities to address higher Bloom's Taxonomy skills.

In addition to providing summative data, use tests to identify intervention needs through script analysis or student interviews. Numerical data may be required for student tracking, but do not provide sufficient detail for good insight into students' strengths, areas for improvement and intervention. Remember that flightpaths are a statistical convenience for schools: students' average attainment rarely follows a smooth trajectory and their actual attainment is usually 'spiky'. However, assessment for learning opportunities take place throughout lessons and provide a rich source of information upon which to plan future lessons and intervention activities.

References

Commission on Assessment without Levels (2015) *Final Report*. UK: Government

Harrison, C. (2016) 'Fostering classroom assessment in science: the importance of transitions', *School Science Review*, **98,** (362), 66–73

Hillier, J. (2012) 'Assessing Pupils' Progress: taking the debate further', *Science Teacher Education*, (64), 28–39

Ofsted (2013) *Maintaining curiosity in science*. Report No. 130135. UK: Government

Stutchbury, K. (2011) 'Assessing Pupils' Progress: issues and opportunities in secondary science and the implications for teacher educators', *Science Teacher Education*, (61), 43–53

CHAPTER 20

International assessments of science: key points explained

Yasmine El Masri

International assessment data are often quoted in discussions about education, but what are these assessments? What are they designed to measure and how might their data be used? In this chapter, the different international assessments are described and the country rankings discussed, along with the limitations of the data.

Introduction

International large scale assessments (ILSAs), such as the Trends in International Mathematics and Science Study (TIMSS), the Progress in International Reading Literacy Study (PIRLS) and the Programme for International Student Assessment (PISA), have become ever more popular in the past decade, with over fifty economies participating in each. Much of the rise in demand has been attributed to the growing pressures for evidence-based educational reforms. ILSAs enable decision-makers to monitor performance trends over time and compare students' performance across countries, as well as benchmark national performance against an international average (IEA, 2016; OECD, 2016a).

The underlying rationale for country comparisons has been arguably to encourage countries to share effective educational policies and learn from each other's strengths and weaknesses. A phenomenon referred to as 'PISA tourism' (Fuhrmann & Beckmann-Dierkes, 2011) has been observed, with many educators visiting top performing countries such as Finland and Shanghai to get exposure to educational practices in those systems. Exchange in instructional methods and teachers' pedagogical content knowledge across countries is certainly valuable and can *theoretically* contribute to advancing the field more globally. Nevertheless, ILSAs' effects, and especially those of PISA, have not been restricted to teachers' knowledge exchange only. In some countries, such as Germany and Norway, the results of PISA 2000 and 2003 led to what has been widely termed the 'PISA Shock', leading to substantial educational reforms in both countries (Elstad, Nortvedt & Turmo, 2009; Ertl,

2006; Steiner-Khamsi, 2003). In England, PISA has been recently adopted as a benchmark for the General Certificate of Secondary Education (GCSE) examinations taken by 16 year-olds (Baird et al, 2016), following a law requiring the regulator to ensure that the standards of English examinations are benchmarked against international standards (House of Commons, 2009).

While few would challenge the value in building educational reforms on solid empirical evidence, ILSAs results have often been misinterpreted and misused, with the discourse being primarily focused on country rankings rather than how data from these tests could be of use to practitioners in the classroom (see Hopfenbeck & Görgen, 2017). The previous edition of *ASE Guide to Secondary Science Education* traced England's performance in Grade 8 TIMSS and PISA since their inception in 1995 and 2000 respectively. This chapter is an update, with a particular focus on implications for science practitioners. England is used as a case study to discuss the data from these ILSAs, because England participated in both the most recent TIMSS and PISA: the process of interpreting these ILSAs that is described here is applicable to the rest of the United Kingdom and, indeed, the rest of the world, although the implications will be specific to each country's context.

Science ILSAs: an overview

TIMSS and PISA are the most prominent ILSAs. They both assess students' knowledge, skills and attitudes in science, amongst other subjects, at different year levels, with some variations between the two assessments. Both also collect extensive background data about students, parents and schools for a more complete picture of students' learning environments.

Trends in International Mathematics and Science Study (TIMSS)

TIMSS is a paper-based assessment, overseen by the International Association for the Evaluation of Educational Achievement (IEA) and administered every four years to Year 4 and Year 8 students (Year 5, age 10, and Year 9, age 14, in England). TIMSS assesses students' achievements in mathematics and science and is tailored to the National Curriculum.

Science assessment framework of Grade 8 TIMSS 2015

TIMSS science is balanced across two dimensions, as shown in Figure 1 (Jones, Wheeler & Centurino, 2013):

- ➤ a content dimension depicting the science knowledge in four content domains – biology, chemistry, physics and earth science; and
- ➤ a cognitive dimension, which captures thinking processes in three domains: knowing, applying and reasoning.

Figure 1: Science assessment framework of Grade 8 TIMSS

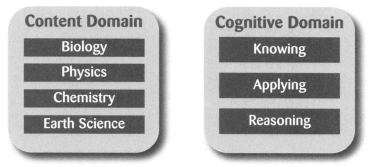

Programme for International Student Assessment (PISA)

PISA is administered every three years to 15 year-olds by the Organisation for Economic Co-operation and Development (OECD). Unlike TIMSS, PISA is claimed to be curriculum-independent; it evaluates what students can do using the knowledge and skills that they have acquired by the end of compulsory education (OECD, 2016a). The survey assesses students' literacy in reading, mathematics and science, with financial literacy as an optional subject and collaborative problem-solving being introduced in 2015. In each cycle, one of the compulsory subjects is assessed as a major subject. In the first cycle in 2000, reading was the major subject; in 2003 it was mathematics; and for 2006 it was science. In 2015, science was again a major subject.

Science assessment framework of PISA 2015

OECD defines science literacy as *'the ability to engage with science-related issues, and with the ideas of science, as a reflective citizen'* (OECD, 2016b, p.20). This definition underlies PISA's science assessment framework, which comprises three inter-related sub-domains: competency, knowledge and system (see Figure 2):

Figure 2: Science sub-domains in PISA 2015

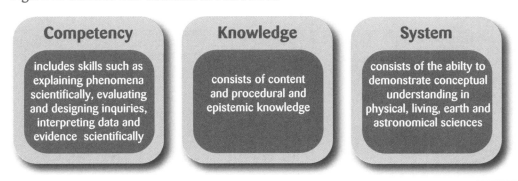

According to OECD (2016b), the three competencies upon which scientific literacy rests are highly dependent on different types of knowledge:

➤ content knowledge: understanding of what is known in different sciences or systems (physical, living, earth and astronomical sciences);

➤ procedural knowledge: understanding of standard procedures used to establish scientific knowledge; and

➤ epistemic knowledge: understanding of the nature and status of scientific knowledge.

While OECD claims that PISA is not curriculum-based, like TIMSS, PISA assesses students' understanding of the traditional school sciences, which are referred to as 'systems' in PISA's science framework (physical, living, earth and astronomical sciences). Moreover, the skills assessed within TIMSS's cognitive domain (knowing, applying and reasoning) are similar to the skills that PISA includes within its competency and knowledge sub-domains, which assess students' ability to use their knowledge, to apply it to new situations and engage in reasoning tasks. Hence, the frameworks of both assessments overlap in many respects. Nevertheless, the science items derived from each assessment framework differ in spirit. This is likely due to the fact that TIMSS measures students' achievements in school science at the end of secondary school (Jones *et al*, 2013), while PISA assesses young people's literacy in science as one proxy of their *'preparedness for life'* (OECD, 2016b p.18). To better illustrate this point, TIMSS- and PISA-released science items are discussed next.

Examples of released TIMSS and PISA science items

TIMSS science items

Figure 3 shows two released Grade 8 science items that target the content domain *earth science*, and the cognitive domain *knowing*. The topic area is *Earth's Processes, Cycles, and History*.

Figure 3a: TIMSS Grade 8 science item (SO4241). (Source: IEA, 2013 p.75)

Where are the active volcanoes most likely to be found?
a. where rivers form
b. where tectonic plates meet
c. where oceans are deepest
d. where land and water meet

Figure 3b: TIMSS Grade 8 science item (SO32126). (Source: IEA, 2013 p.22)

> **State one way that a volcanic eruption can affect the environment**

Figure 3a shows a multiple-choice item with 'B. *where tectonic plates meet*' a correct response. Figure 3b illustrates a short constructed-response item, and the correct response is determined using a scoring guide that includes criteria and examples for correct and incorrect responses (see Appendix 1).

Both TIMSS items require content knowledge of how volcanoes form and where they are mainly found. Neither of these items includes stimulus material (e.g. text, photo, map, etc.) so students' responses are based on the content that they have already learned in school. This is not to suggest that TIMSS items never include any visual representations or text.

PISA science item

Science items in PISA are organised in units; each unit covers a particular topic and consists of at least one stimulus (text, graphs, photo, etc.) and several questions. Figure 4 (overleaf) shows a released PISA science item with one out of the four questions in the unit 'Volcanic eruptions' listed on the left and stimulus material appearing on the right.

The item targets the competency *Interpret Data and Evidence Scientifically*, the *Procedural* knowledge/system and the *Global – Hazards* context, and is considered to place low cognitive demands on the students. It is a multiple-choice item with map location *D* being the correct answer. This item, as with all other multiple-choice items in PISA 2015, has been computer-scored[1].

Responding to TIMSS items requires some prior knowledge of plate tectonics and volcanic activity, whereas answering PISA items relies more on the ability to draw information from the stimulus material. Arguably, students should be able to answer this question without having covered the topic at school. Granted that this is the case, the item is likely to place higher cognitive demands on students who have not learned this topic in school (see Nardi, 2008). In England, students learn about plate tectonics and volcanic eruptions in geography, first at Key Stage 2 (ages 7-11) under 'physical geography' and again at GCSE level. With geography not being a compulsory subject at GCSE, some students will have only covered this topic in primary school, and will probably be at a disadvantage compared to students elsewhere in the world who have covered this topic more recently in their curricula. This raises questions around the extent to which comparing students across countries is a fair comparison when they have not all been exposed to the same curriculum (El Masri, Baird & Graesser, 2016; Huang, Wilson & Wang, 2016).

Figure 4: PISA science item (CS644Q01)[2]. (Source: OECD, n.d.)

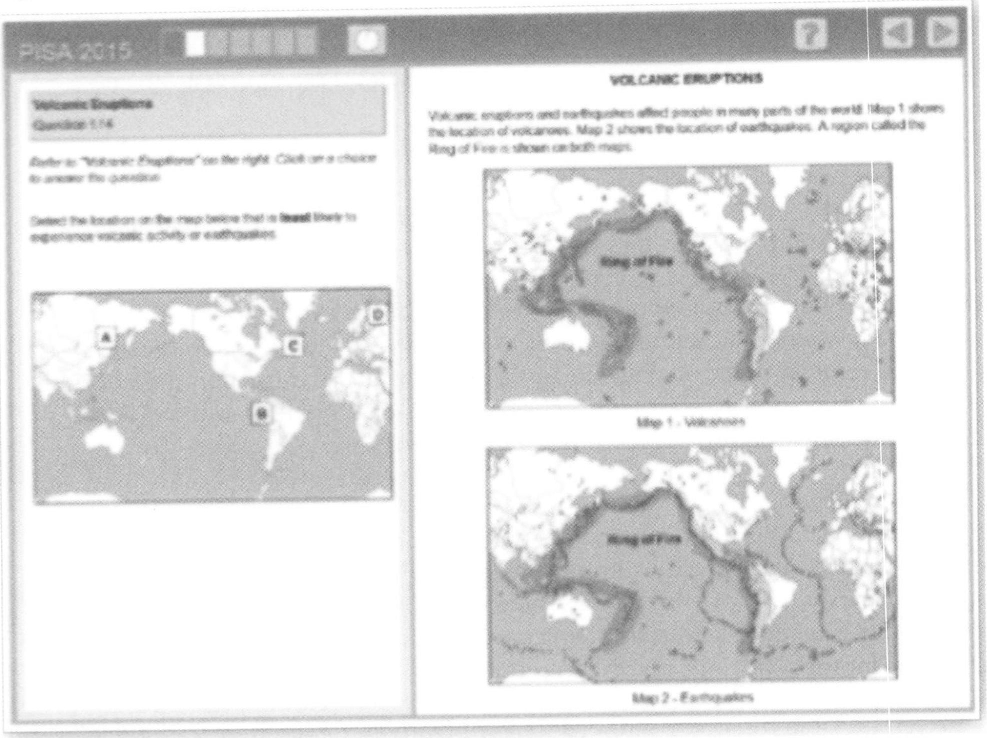

Similarity with GCSE science items

The items in Figure 5a and 5b are taken from a GCSE geography paper administered in England covering the same topic (volcanic eruptions) as the sample TIMSS and PISA items discussed earlier.

Familiarity of the format and type of items can influence demands placed on students when solving questions (see El Masri, Ferrara, Foltz & Baird, 2017). Like PISA items, GCSE items are organised in units where several questions relate to the same stimulus, and Item 1(a) is extracted from the unit 'The Restless Earth'. Answering questions in this unit relies to a large extent on drawing information from the map (see Appendix 2). This item is very similar to the PISA item discussed earlier, except that it is not computer-based. Item 1(b) relies more on knowledge acquired in the classroom. It is similar to the TIMSS item in Figure 3b, in the sense that it is a constructed-response item that is

[1]Constructed-response items are human-scored following a guide similar to the one of TIMSS (see Appendix 1).

[2]PISA 2015-released field trial generic items: https://www.oecd.org/pisa/test/PISA2015-Released-FT-Cognitive-Items.pdf

Figure 5a: GCSE item 1(a) (90301H). (Source: AQA, June 2016)

1	**The Restless Earth** **Total for this question: 25 marks**

1   **The Restless Earth**   **Total for this question: 25 marks**

1 (a)   Study Figure 1 on the insert, an atlas map showing the Earth's tectonic plates.

Use Figure 1 to complete the Fact File.

(4 marks)

> Fact File
>
> Names of plates at X on the plate margin ....
>
> plate and .... plate
>
> Name of volcano at 43°N 122°W..
>
> Distribution of tectonic features in South America..
>
> ....
>
> ....

Figure 5b: GCSE item 1(b) (90301H). (Source: AQA, June 2016)

1 (b)   Explain the formation of fold mountains

(4 marks)

curriculum-dependent. However, the TIMSS item requires a shorter answer and was worth only 1 mark (as opposed to 4 marks allocated to the GCSE item). The marking scheme for item 1(b) provides elements of a correct response, with two broad marking categories (see Appendix 3).

England's performance in science ILSAs in 2015

England ranked amongst the top 10 highest performers in Grade 8 TIMSS and the top 15 in PISA, in 2015.

TIMSS 2015: England's performance in Grade 8 science

4,814 Year 9 students randomly selected across 143 secondary schools in England participated in Grade 8 TIMSS (Greany, Barnes & Pensiero, 2016). England's performance in Grade 8 TIMSS has been consistent in science since the assessment's first cycle in 1995, achieving 537 in 2015, significantly higher than the international mean[3] (see Figure 6 overleaf).

[3]TIMSS international scale has a mean of 500 and a standard deviation of 100.

Figure 6: Country grouping based on relative performance to England's in science in Grade 8 TIMSS in 2015. (Source: IEA, 2016)
(Note: Neither Wales nor Scotland participated in TIMSS 2015, and Northern Ireland only participated in the Year 6 (age 11) assessment.)

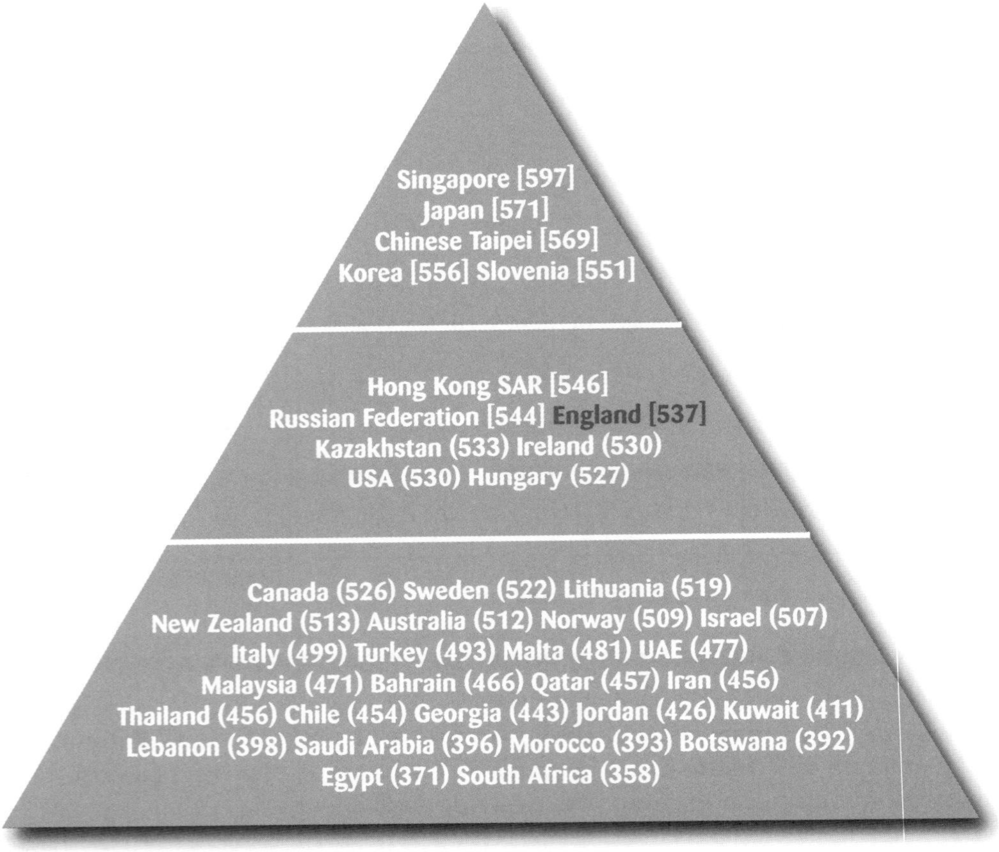

Singapore [597]
Japan [571]
Chinese Taipei [569]
Korea [556] Slovenia [551]

Hong Kong SAR [546]
Russian Federation [544] England [537]
Kazakhstan (533) Ireland (530)
USA (530) Hungary (527)

Canada (526) Sweden (522) Lithuania (519)
New Zealand (513) Australia (512) Norway (509) Israel (507)
Italy (499) Turkey (493) Malta (481) UAE (477)
Malaysia (471) Bahrain (466) Qatar (457) Iran (456)
Thailand (456) Chile (454) Georgia (443) Jordan (426) Kuwait (411)
Lebanon (398) Saudi Arabia (396) Morocco (393) Botswana (392)
Egypt (371) South Africa (358)

At domain level, England performed above the international mean, with a weaker performance in chemistry, which is probably due to the fact that one out of the 22 science topics included in TIMSS 2015 (the role of electrons in chemical bonds) was taught *after* Year 9 in England (Greany et al, 2016). While the results of Grade 8 TIMSS in science have been static over time in England, Year 9 students demonstrated an improvement in this subject in 2015 as compared to their performance four years earlier in Grade 4 TIMSS 2011.

PISA 2015: England's performance in science
5,194 fifteen year-old students, randomly selected from 206 schools in England, participated in PISA 2015 (Jerrim & Shure, 2016). England ranked amongst the top 15 highest performing countries in science in 2015, with an average score

Figure 7: Country ranking based on overall performance in science in PISA 2015[4]. (Source: Jerrim & Shure, 2016)

	Country	Mean
Countries more than 20 points ahead of England	Singapore	556*
	Japan	538*
	Estonia	534*
	Taiwan	532*
Countries between 10 and 20 points ahead of England	Finland	531*
	Macao	529*
	Canada	528*
	Vietnam	525*
	Hong Kong	523*
Countries within 10 points of England	China	518
	South Korea	516
	New Zealand	513
	Slovenia	513
	England	**512**
	Australia	510
	Germany	509
	Netherlands	509
	Switzerland	506
	Ireland	503*
Countries between 10 and 20 points behind England	Belgium	502*
	Denmark	502*
	Poland	501*
	Portugal	501*
	Northern Ireland	**500***
	Norway	498*
	Scotland	**497***
	United States	496*
	Austria	495*
	France	495*
	Sweden	493*
	Czech Republic	493*
	Spain	493*
Countries between 20 and 30 points behind England	Latvia	490*
	Russia	487*
	Wales	**485***
	Luxembourg	483*

Legend:

Significantly above OECD average

Not significantly different from the OECD average

Significantly below OECD average

OECD average = 493

*Mean score is significantly different from England's at the 5% level

[4]The table does not include countries with a mean score in science more than 30 points below England's mean score.

significantly higher than the OECD average (see Figure 7 p.253), consistent with its performance since 2006. Consistency has been also observed across the different sub-domains, with very little variation across competencies, knowledge types and systems (Jerrim & Shure, 2016).

Beyond country ranking: implications of TIMSS and PISA for science teachers

England's solid performance in science in TIMSS and PISA over the years could be considered as a testimony to the high quality standards of the English educational system, with both assessments providing empirical evidence that, by the end of compulsory education, students acquire the scientific knowledge and skills necessary for them to be proactive and productive members of the society.

The high quality standards are reflected in the positive attitudes towards science that Year 9 students reported in the 2015 cycle of TIMSS and PISA. Both studies suggest that secondary students in England enjoyed science and valued science lessons (Greany et al, 2016; Jerrim & Shure, 2016). These attitudes were not necessarily shared by their peers in the top performing countries in East Asia (Greany et al, 2016). PISA data suggest that science classrooms in England are more interactive than in highly performing countries (Jerrim & Shure, 2016), which could also partly explain students' positive attitudes towards science lessons. In addition, TIMSS data suggest that students in England felt confident in science (Greany et al, 2016). Indeed, data from PISA 2015 questionnaires suggest that a larger proportion (~28%) of 15 year-olds in England consider science-related careers, as opposed to 24% across developed countries (Jerrim & Shure, 2016).

Based on both TIMSS and PISA questionnaire data, secondary students in England go to well-resourced schools with students having wide access to computers and equipment in science classes (Greany et al, 2016; Jerrim & Shure, 2016). TIMSS data suggest that there are no major behaviour or safety concerns in schools in England; however, in PISA, a significantly larger proportion of students in England reported frequent occurrences of low level disruption in class (e.g. students not listening to the teacher, noise and disorder in the class, etc.) than the OECD average or their peers in the highest performing countries. This could potentially be associated with the interactive nature of science classes in England; however, excess disruption could negatively affect the learning environment of students.

TIMSS and PISA data also point to challenges in the system. Headteachers reported difficulties in recruiting science teachers, with issues associated with staff absenteeism and the need to better train teachers in differentiated learning (Greany *et al*, 2016; Jerrim & Shure, 2016). Year 9 science teachers scored low on the job satisfaction scale in TIMSS, with many teachers identifying teaching load as well as the frequent curricular changes as main areas of concern (Greany *et al*, 2016).

Shortages of science teachers and job satisfaction are not the only challenges in the science education system in England. While the gender gap in science has closed in both Grade 8 TIMSS and PISA in 2015, both assessments point to a wide socio-economic gap separating the advantaged and the disadvantaged in England (Greany *et al*, 2016; Jerrim & Shure, 2016). This gap also exists between students for whom English is a first language and those with English as an additional language (EAL). PISA data also suggest that the achievement gap is observed between students born and raised in England and students of immigrant backgrounds, and between white students and black or Asian students. Moreover, PISA data suggest that students attending schools rated as outstanding by Ofsted are up to two years ahead, in science, of their peers in schools rated as inadequate. These data are invaluable for planning educational reforms that are more inclusive and which provide access to quality science education for everyone in England.

Limitations of the data

TIMSS and PISA provide the public, educators and policymakers with extensive data about the national science education system, including factors that are normally associated with students' achievements in science (e.g. students' attitudes, socio-economic status, resources, etc.). These data can be very informative for all stakeholders and can form a robust basis for building and improving science teaching and learning, preparing well-informed citizens who are armed with the necessary knowledge and skills to contribute to society. However, ILSAs have been strongly criticised in academia on many grounds (see Hopfenbeck *et al*, 2017; Lindblad, Pettersson & Popkewitz, 2015). The following section outlines the limitations most relevant here:

Country rankings
There has been heavy criticism of the focus on country rankings, with many researchers questioning their reliability on technical grounds (e.g. Kreiner & Christensen, 2014). Country rankings vary based on the number and selection of countries participating in the assessments and hence should not be taken as a yardstick or evidence of systemic improvement or regression.

International comparisons
While there is certainly some value in comparing how secondary students perform in ILSAs across countries, there has been strong criticism of the extent

to which it is possible to directly compare outcomes across countries, especially across those that have extreme variations in language, culture, curriculum priorities and wealth (Ercikan & Koh, 2005; Grisay et al, 2007; Nardi, 2008). While TIMSS and PISA follow rigorous methods of translation and adaptation of tests in over 40 languages, and while PISA claims to be independent of national curricula, empirical evidence shows that items do not behave the same way across countries, with curriculum and language being key sources of bias across language versions of the same science test (El Masri et al, 2016; Huang et al, 2016).

Over-interpreting or misinterpreting data

ILSAs results should be interpreted at population level and not at individual student or school level. Samples are formed to be representative of the target student population at national level and hence it is misleading to apply conclusions to individual schools or individuals (Jerrim & Shure, 2016). On a similar note, ILSA outcomes cannot be used as direct indicators of school quality and effectiveness, because scores on these assessments are the result of various factors in addition to the schools that students attended (e.g. socio-economic status, parents' support, attitudes and motivation, etc.). Likewise, the variation of PISA or TIMSS average scores cannot be attributed to specific education reform, as a host of other factors, such as the variation in economic conditions or an increase in the proportion of immigrants, also affect these results.

In conclusion

This chapter offers an overview of two science ILSAs, TIMSS and PISA, and what these assessments can infer about science teaching and learning in England. Both PISA and TIMSS data suggest a solid profile for England in science at the end of compulsory education (Greany et al, 2016; Jerrim & Shure, 2016). Students reported enjoying science and recognising its value in preparing them for their future careers. Schools have been reported to be well resourced, with good access to computers in the science classroom.

While both tests point to a practically non-existent gap in science achievement between males and females, they both indicate relatively large socio-economic and ethnic gaps in England that appear to be wider than in many OECD countries. In addition, the shortage of science teachers appears to be another source of concern in secondary schools, which would be worth addressing in future planning.

In this chapter, some limitations of ILSAs have been highlighted in the hope that this will provide teachers with the necessary understanding with which to engage with ILSAs results and take part in the wider policy debate. Baird *et al* (2016) suggest that educational policies are still, as always, mainly motivated by ideology, which policymakers often mask using empirical evidence from ILSAs. It is therefore crucial to support stakeholders, including teachers, to better understand ILSA data and their limitations, to help them critically evaluate ILSA evidence, uncover real incentives behind reforms and contribute positively to policy discussions.

References

Assessment and Qualification Alliance (n.d.) *AQA Past Papers. GCSE Geography 90301 Higher Tier June 2016*. Retrieved on 23rd June 2017 from: http://filestore.aqa.org.uk/sample-papers-and-mark-schemes/2016/june/AQA-90301H-QP-JUN16.PDF

Assessment and Qualification Alliance (n.d.) *AQA Past Papers: GCSE Geography 90301 Higher Tier June 2016 Marking Scheme*. Retrieved on 23rd June 2017 from: http://filestore.aqa.org.uk/sample-papers-and-mark-schemes/2016/june/AQA-90301H-W-MS-JUN16.PDF

Baird, J.-A., Johnson, S., Hopfenbeck, T.N., Isaacs, T., Sprague, T., Stobart, G. & Yu, G. (2016) 'On the supranational spell of PISA in policy', *Educational Research*, **58,** (2), 121–138. http://doi.org/10.1080/00131881.2016.1165410

El Masri, Y., Ferrara, S., Foltz, P. & Baird, J. (2017) 'Predicting item difficulty of science national curriculum tests: the case of key stage 2 assessments', *Curriculum Journal*, **28,** (1), 59–82. http://doi.org/10.1080/09585176.2016.1232201

El Masri, Y.H., Baird, J. & Graesser, A.C. (2016) 'Language effects in international testing: The case of PISA 2006 science items', *Assessment in Education: Principles, Policy & Practice*, **23,** (4), 427–255. http://doi.org/http://dx.doi.org/10.1080/0969594X.2016.1218323

Elstad, E., Nortvedt, G.A. & Turmo, A. (2009) 'The Norwegian Assessment System: An Accountability perspective', *CADMO*, **17,** (1), 89–103

Ercikan, K. & Koh, K. (2005) 'Examining the construct comparability of the English and French versions of TIMSS', *International Journal of Testing*, **5,** (1), 23–35

Ertl, H. (2006) 'Educational standards and the changing discourse on education: The reception and consequences of the PISA study in Germany', *Oxford Review of Education*, (32), 619–634

Fuhrmann, J.C. & Beckmann-Dierkes, N. (2011) 'Finland's PISA success: Myth and transferability', *KAS International Reports*

Greany, T., Barnes, I. & Pensiero, N. (2016) *Trends in Maths and Science Study (TIMSS): National Report for England, (November)*

Grisay, A., de Jong, J.H.A.L., Gebhardt, E., Berezner, A. & Halleux-Monseur, B. (2007) 'Translation equivalence across PISA countries', *Journal of Applied Measurement*, **8**, (3), 249–266

Hopfenbeck, T.N., Lenkeit, J., El Masri, Y.H., Cantrell, K., Ryan, J. & Baird, J.J-A. (2017) 'Lessons learned from PISA: A systematic review of peer-reviewed articles on the Programme for International Student Assessment', *Scandinavian Journal of Educational Research.* http://doi.org/http://dx.doi.org/10.1080/00313831.2016.1258726

House of Commons (2009) *Apprenticeship, Skills, Children and Learning Bill. Bill 55*

Huang, X., Wilson, M. & Wang, L. (2016) 'Exploring plausible causes of differential item functioning in the PISA science assessment: language, curriculum or culture', *Educational Psychology*, **36**, (2), 378–390. http://doi.org/10.1080/01443410.2014.946890

IEA (2016) *IEA's Trends in International Mathematics and Science Study - TIMSS 2015.* Retrieved from http://timss2015.org/download-center/

International Association for the Evaluation of Educational Achievement (2013) *TIMSS 2011 user guide for the international database.* TIMSS & PIRLS International Study Center, Lynch School of Education, Boston College

Jerrim, J. & Shure, N. (2016) *Achievement of 15 year-olds in England: PISA 2015 National Report.* London: Department for Education

Jones, L.R., Wheeler, G. & Centurino, V.A.S. (2013) 'TIMSS 2015 science framework'. In *TIMSS 2015 Assessment Frameworks,* Mullis, I.V.S. & Martin, M.O. (Eds.), pps. 29–59. Chestnut Hill, MA: TIMSS & PIRLS International Study Center, Boston College

Kreiner, S. & Christensen, K.B. (2014) 'Analyses of model fit and robustness. A new look at the PISA scaling model underlying ranking of countries according to reading literacy', *Psychometrika*, **79**, (2), 210–231

Lindblad, S., Pettersson, D. & Popkewitz, T. (2015) *International comparisons of school results – A systematic review of research on large scale assessments in education.* Stockholm, Sweden: Swedish Research Council

Nardi, E. (2008) 'Cultural biases: A non-Anglophone perspective', *Assessment in Education: Principles, Policy & Practice*, **15**, (3), 259–266

OECD (n.d.) *PISA 2015 released field trial generic items. OECD.* Retrieved from https://www.oecd.org/pisa/test/PISA2015-Released-FT-Cognitive-Items.pdf

OECD (2016a) *PISA 2015 Results (Volume 1): Excellence and equity in education*. Paris: OECD Publishing

OECD (2016b) 'PISA 2015 Science Framework'. In *PISA 2015 Assessment and Analytical Framework: Science, Reading, Mathematic and Financial Literacy* (pps. 17–46). Paris: OECD Publishing. http://doi.org/http://dx.doi.org/10.1787/9789264255425-3-en

Steiner-Khamsi, G. (2003) 'The politics of league tables', *Journal of Social Science Education*, (1), 1–6

Acknowledgements

Special thanks to Associate Professor Therese N. Hopfenbeck and Associate Professor Judith Hillier for their invaluable comments on earlier drafts of this chapter.

Appendix 1: TIMSS scoring guide for item S032126). (Source: IEA, 2013 p.23)

Code	Response	Item: S032126
	Correct Response	
10	States the **negative** environmental effect due to volcanic eruptions such as pollution (due to release of gases, smoke, ash, etc.) or destruction of habitats or plant/animal life (due to lava flow, burning or similar). Examples: Burns away essential plant life. Lava would ruin the ground and burn everything. It lets out harmful gases. It covers everything in its path. [Assume 'in its path' means lava flow.] Volcanic eruptions produce ashes that will pollute the environment. It will release carbon dioxide into the atmosphere that might cause a greenhouse effect. The huge amount of black smoke will pollute the air. Sulfuric gases cause acid rain.	
11	States a **positive** environmental effect such as making land fertile, creating new habitats, and allowing for different life forms. Examples: It can make the land surrounding the volcano more fertile. It might destroy some crops, but give a better chance for a new one to come in.	
19	Other correct [responses]	
	Incorrect Response	
70	Gives only a general statement of destruction or the nature of volcanic eruptions with inadequate description of how the environment is affected. Examples: It can destroy everything. People can die. It can ruin the environment. It is very hot and the heat might get out and affect the environment. Dense ash and lava.	
79	Other incorrect [responses] (including crossed out, erased, stray marks, illegible, or off task)	
	Non-response	
99	Blank	

Appendix 2: Figure 1 of GCSE question 90301H. (Source: AQA, June 2016)

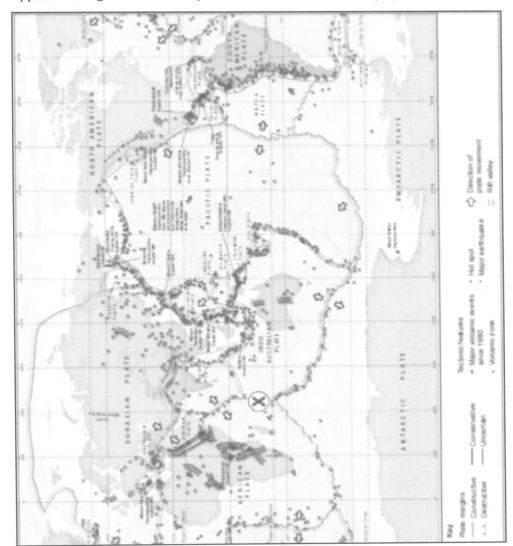

Appendix 3: Marking scheme AQA GCSE item 1a and 1b (90301H)[5].
(Source: AQA, June 2016)

Question 1: The Restless Earth

1(a) C   **Fact File**
(4 marks)
AO2 – 2
AO3 – 2

Name of plates at X on the plate margin	African and Indo-Australian
Name pf volcano at 43°N 122°W	Mt St Helens
Distribution of tectonic features in South America	The western area has many earthquakes – extending right along the coast; some volcanoes occur in the same areas, such as the north western area near the Equator and in a line following the west coast. These occur in a fold mountain range – the Andes at a (destructive) plate boundary.

1 mark for each of the first two parts and 2 x 1 for simple points or 1+1 for an elaborated point for the final part

1(b) C   Fold mountains are formed from layers of sedimentary rock built up   (4 marks)
over millions of years. Rivers transport material to the oceans/
geosyclines and sediment falls to the bed of the ocean and is   AO1 – 4
deposited there. The weight of the water and subsequent layers of
sediment leads to copaction and cementation as the layers are formed.
As the plates move together, the rock layers are compressed/crumpled
to form high areas known as anticlines and lower areas are known as
synclines. There may be reference to formation at subduction zones as
well as collision as subtypes of destructive plate margins.

Level 1 Basic (1 – 2 marks)
Partial sequence – may address formation of sedimentary rock or
plate movement only
Statements are general and separate in random order

Level 2 Clear (3 – 4 marks)
There is a clear reference to both aspects – recognition of sedimentary
rock formation and plate movement – in a more complete sequence

[5] Marking scheme available at http://filestore.aqa.org.uk/sample-papers-and-mark-schemes/2016/june/AQA-90301H-QP-JUN16.PDF

SECTION 5:

Science teaching as a profession

CHAPTER 21

Developing professional practice

Alex Manning & Emma Towers

Much of science initial teacher education (ITE) necessarily focuses on developing professional and pedagogical skills. This chapter aims to highlight some of the issues teachers may encounter after their ITE year upon which they can reflect as they continue to develop their professional practice.

Introduction

An ITE programme will no doubt set trainees up well to take on the challenges and rewards of becoming a science teacher and what to expect in their first year of teaching. However, these programmes do not have time or scope to explore in much detail what happens after the ITE year, or how to sustain a commitment to teaching. In this chapter, we consider motivations for becoming a teacher before providing a background to the teacher retention issues in England. We examine how beginning teachers might develop resilience in order to sustain them in their career and offer some ways to consider how they might find the right 'fit' in a school or department. Finally, we discuss the teaching journey and offer some ways to consider how beginning teachers might navigate their own teaching career so that they make informed decisions about the direction of that career.

Motivations for becoming a teacher

There are myriad reasons why someone chooses to teach, as demonstrated by a substantial body of international literature (for example, Fokkens-Bruinsma & Canrinus, 2012; Heinz, 2013; Watt & Richardson, 2012). Common reasons that teachers enter the profession include enjoyment of working with young people, the desire to make a difference to children's lives, or the wish to promote a particular subject.

Findings from teacher motivation studies are useful in informing an educational recruitment and retention policy, for two key reasons. Firstly, they can provide a comprehensive overview of beginning teachers' motivations, therefore giving insight into effective design of recruitment strategies. Secondly, they help to identify what aspects of the teaching profession provide job satisfaction, which may influence whether people are likely to stay in the profession. Ultimately, understanding beginning teachers' motivations not only provides an appreciation of the factors that attract people to teaching, but also may influence teachers' subsequent career decisions. At a time when teacher retention rates are of concern in England, it is worthwhile exploring our initial motivations for deciding to teach, as these may help with subsequent career decisions, such as whether one decides to stay in or leave a current position, whether one moves to another school, or adapts one's role within the school.

Background to retention

To place the issue of teacher retention in context, it is valuable to provide a brief overview of the teaching workforce landscape in England. Rising student numbers, together with a growing shortage of teaching recruits and concerns over the numbers of teachers who report that they are considering leaving the profession, have led to a crisis in teacher recruitment and retention. Recruitment and retention are two distinct issues in teacher supply; recruitment is about attracting individuals into the profession, while retention focuses on keeping those same individuals in their teaching positions. At the same time that recruitment numbers fail to hit targets, high numbers of recently qualified teachers are leaving the profession. Recruiting and educating teachers is both time-consuming and expensive if those individuals do not continue to stay in the classroom.

Figures on the school workforce in England (DfE, 2016) show that teacher retention rates reduce after each year of qualifying. After five years, 70% of qualified teachers remain in their posts, after ten years this reduces to 61% and just 50% of teachers remain in service 19 years after qualifying and entering the profession (DfE, 2016). These figures do not take into account the variation between the retention rates in primary and secondary schools, different locations such as rural and urban areas, and different types of schools. More importantly for us, these figures do not acknowledge the difference in retention rates for different subjects in the secondary phase. Indeed, science and maths teachers account for the highest rates of leaving the profession (Worth & De Lazzari, 2017). Therefore, a science teacher is particularly valuable and sought after.

It is not easy to gain an accurate picture of the teacher recruitment and retention issue, in part because there is a wide array of often contradictory data sources, making it difficult to appreciate the complexity of the situation. What is clear, however, is that teacher recruitment and retention in England is deeply concerning to many.

Developing resilience and wellbeing

Above we have outlined the issues in England regarding teacher retention. However, in this chapter, we are concerned with how to retain science teachers in the science teaching profession. Research shows that teacher wellbeing is linked to pupil wellbeing and educational outcomes (Roffey, 2012; Briner & Dewberry, 2007). It seems obvious that a healthy teacher with high job satisfaction and a positive attitude to his/her work and students is more likely to perform well. Therefore, attending to teachers' sense of wellbeing is crucial if schools are to retain happy healthy teachers. Although the school structure and culture will be central in either promoting teacher wellbeing or contributing to teacher stress and burnout, there are a number of key things that a teacher can do for him/herself. Teaching is a demanding profession, so it is essential that a teacher takes some time for 'me' during each working day, e.g. taking time in the day to take a complete break to eat lunch, preferably away from the classroom. With the pressure to get on top of heavy workloads, many teachers find themselves marking work, or preparing for a lesson during their lunchtime. Whilst on occasion this may be unavoidable, it is important that we try to make time, as far as possible, to take a break for at least some part of the lunch break. Sleep studies are often quoted in the media, but getting enough quality sleep is crucial if we are to arrive fresh and ready for a class of students each day (Willis, 2014). This means ensuring that work does not encroach on the nighttime routine and that we are not marking books well into the night. Similarly, a fit teacher is a healthy teacher. There is a clear link between poor physical health and stress (APA, 2013), so it is important to take time to exercise regularly during the week. Many schools now provide wellbeing activities such as yoga, pilates and mindfulness classes, which can help improve physical and mental wellbeing, and also may help build relationships with other members of staff beyond one's usual contacts. Many experienced teachers will say that school life can be all-consuming, infiltrating all aspects of life. As such, we would encourage teachers to make time for hobbies and friends and try as much as possible to maintain boundaries between home and work, the working week and the weekends, in order to maintain a healthy work-life balance.

Wellbeing is interlinked with resilience and this concept of resilience is a useful one to explore in helping us to understand how to retain science teachers. While we acknowledge that, for some, a resilient act might be to leave the teaching profession, here we are concerned with developing the resilience to sustain the science teacher role.

Resilience was once viewed as being an innate and stable quality. It is now understood to be more fluid and therefore malleable. Gu and Day (2007, p.1302) suggest that resilience indicates that a teacher has the capacity to 'recover strengths or spirit quickly and efficiently in the face of adversity', asserting that resilience is best understood as 'a relative, multi-dimensional and developmental construct' (Gu & Day, 2013 p.25). There are multiple influences on a teacher's resilience, which shift and change according to the teacher's personal and professional contexts and circumstances (Day & Gu, 2010). Personal characteristics that enable resilient responses to challenging situations include a strong sense of self-belief, self-esteem and a sense of humour. While personal qualities and attributes are considered to be important influences on a teacher's resilience, they are not the sole factors that determine a teacher's level of resilience.

Another factor considered to influence teacher resilience are the relationships that teachers build and sustain with colleagues and school leaders. In fact, Day and Gu (2010) argue that 'relational' resilience is the most important factor contributing to whether a teacher stays in or leaves the job. Here we can see a clear link between a teacher's sense of wellbeing and his/her capacity for resilience. Certainly, the quality of relationships built and sustained in a school has a significant impact on a teacher's wellbeing, as well as the ability to cope with the stresses and demands of daily school life. Positive relationships with colleagues and school leaders can boost teachers' sense of resilience, as can feeling supported through well-structured and organised workplaces. Similarly, positive relationships with students not only enhance a teacher's sense of wellbeing, but also impact positively on students' own wellbeing and their academic success.

It may be that resilience waxes and wanes throughout a career as those factors that enhance or reduce resilience also change and shift. The concept of 'relational' resilience is an important one for us and we should consider ways to develop our own community of practice (Lave & Wenger, 1998). Collaborative strategies you might adopt include: taking part in peer observations, peer teaching or planning within your science department; engaging with online forums and communities, attending and/or organising TeachMeets with local science teachers; and/or reconnecting with your ITE peers. This is by no means an exhaustive list, but it does serve to highlight the ways to connect with others and develop 'relational' resilience.

Will you stay? Where do you fit?

Consider the 'paradox' of teaching, where a teacher can be 'in love with one's work, but daily talk of leaving it' (Nias, 1989 p.191). Teaching can be intensely rewarding and uplifting and those who choose to teach are often fuelled by a passion for their subject, the job and a commitment to their students. However, teaching is also a relentless profession, fraught with tensions, difficulties and fatigue. The demands of the role can be exhausting, not least as a result of increased responsibilities; teachers can experience conflicts with other members of staff and have to deal with difficulties with students, such as challenging behaviour. Furthermore, teaching can at times mean boredom and repetition, where teachers are required to work within the confines of internal and external policy directives. It is this 'paradox' of which teachers should take heed when making decisions about their career futures. For many, there comes a 'crossroad' in their careers, where they need to take stock and consider how they have arrived where they are and what they need to do to sustain their motivation and commitment to the profession. We may be considering where we 'fit' into the school and the profession and how this in turn fits into our wider life, including our personal priorities and commitments. One way to consider our 'fit' is to reflect on potential reasons for staying – our 'pull' factors. Similarly, we may wish to reflect on the reasons why we may choose to leave – our 'push' factors (Manning & Towers, in press). It is a valuable exercise to take time to reflect on our motivations; are there key 'pull' factors that are likely to sustain us in our roles? Are there any potential 'push' factors that may need attention? Perhaps these involve feelings about the leadership of the school, colleagues, opportunities for career development, or perhaps relationships with the students? In the next section, we discuss in some more detail what factors could motivate a teacher to remain in post.

The teaching journey

A teaching career can be thought of as a journey that follows a complex travel system. There is no one start or finish point, neither is there a single route between them, but rather multiple personalised career journeys. Do talk to other teachers, listen to their journeys – they will all be different – we can learn a little from each of them. Extending the analogy of the 'teaching journey', here we explore those factors that may influence a decision to stay; we refer to them as 'pit stops' that might be experienced on the journey. The crossroads, as discussed previously, can help in making choices, and deciding which 'pit stops' will enhance a teaching journey and contribute towards a sense of fulfilment in the role, and which may encourage a teacher to remain in the profession. One way to think of these journeys might be as a travel network; not only are there multiple routes on the same network system between two points, there are also different travel systems available between points on the journey. This is a useful mental model to keep in mind when reflecting on a teaching journey.

When considering what we need for career fulfilment, it is useful to remember that teachers' motivations for remaining in their posts can be complex, multi-faceted, and dependent on their individual situations, both personal and professional. There are indeed some common motivations that teachers cite as reasons to stay: having a strong sense of self-belief and the feeling that they are making a difference to students' lives and learning; having a good and supportive leadership team and Headteacher; opportunities for continuing professional development (CPD); and good and collaborative relationships with colleagues. Given that many teachers cite 'making a difference' as a reason to enter the teaching profession, they need to feel that they are indeed making a difference if they are to sustain their commitment.

Reflection

When 'becoming' a science teacher, the phrase 'a reflective practitioner' was no doubt often heard and trainees are often encouraged to reflect on their experiences. This process should be maintained; continually reflecting on who we are, where we are and where we are going will enable us to make better-informed decisions and enhance our job satisfaction. Therefore, seek out opportunities to engage in more self-reflection, either from courses offered externally or by encouraging leadership teams to provide CPD sessions that focus on self-reflection of teachers in the school. This form of professional development is one that can contribute to the *'commitment, resilience and health needs of teachers'* throughout their professional lives (Day, 2012 p.14). Other ways to develop reflectivity might be to keep a reflective journal, talk to others as described above or identify a mentor/coach who can help to articulate our ongoing needs.

Engaging with research literature is a further way to reflect on practice. Science education research is an active field, rich in literature, and reading such materials can prompt us to reflect on our own experiences and practices in school and the classroom. There are three key ways to access research: through academic journals such as *International Journal of Science Education (IJSE)* or *Science Education;* professional journals, for example *School Science Review (SSR)* or *Physics Education;* and, of course, published books. Such research might be read as an individual endeavour or through a 'book club' style, where colleagues/department members select, share and read the same article, then meet to discuss thoughts and implications arising from the reading. Of course, more formal engagement with literature can be achieved through further study/qualifications (discussed below).

Role awareness

Our reflections could prompt us to think about revisiting our current roles within school and possibly changing our professional roles, which may result in strengthening our commitment to our jobs and the profession. We might feel the need to take on a new responsibility, leadership or mentoring, pastoral or curriculum. Conversely, maybe 'downshifting' (Troman & Woods, 2000) would better meet our needs. For example, perhaps we have taken on too much responsibility, leaving us feeling that we are not doing any of our roles well. Perhaps our roles and levels of responsibility are negatively impacting on our work/life balance or we may simply be on a professional track that is not suiting us. If teachers find that an aspect of their job is not going according to plan, they can search for another manner through which they can contribute to their profession and in which they find new career fulfilment and satisfaction (Rinke, 2014). Maintaining an open and honest dialogue with line managers is crucial so that current commitments and career trajectories can be discussed.

Ethos alignment

It is perhaps an obvious statement to make, but the culture and ethos of a school has an impact on teacher retention. Indeed school leaders, in particular Headteachers, are in a large part responsible for creating the culture and ethos of their schools. Therefore, it is perhaps unsurprising that teachers frequently cite leadership as a key factor in their decision to stay in their schools (for example, Hong, 2012). Reading the school's vision and aims on the school website or the most recent Ofsted report may not necessarily provide a new teacher applying to the school with sufficient information about the school's ethos. The vision of the school is best communicated by what can be seen and felt, rather than what is said, or read about in a school policy document, so do try to visit a school as part of the application process.

There are some key points to note when visiting a new school, either prior to applying for a post or on interview day. The school's ethos permeates all aspects of school life. Certainly, meeting the Headteacher and Head of Science (HOS) is crucial for gauging the ethos of the school/department. A Headteacher/HOS who values his/her staff will take time to welcome an applicant to their school and answer questions. A Headteacher showing an interest in an applicant is a good indication of someone who cares about the staff s/he employs. A Headteacher who shows that s/he respects and values the school staff and the students will be more likely to foster an atmosphere of collegiality and friendliness within the school. An open school will welcome an applicant visiting and observing lessons. Notice how teachers and students communicate with each other. Positive teacher-student relationships, as well as the friendliness of the students and their interaction with one another, are telltale signs of a positive school ethos. Additionally, notice if students are engaged in their learning and are being sufficiently challenged; do the teachers appear to have high expectations of the students? Do the students demonstrate good behaviour in and out of the classroom? This will provide a good indication that school organisational structures are understood and valued by everyone. Finally, spend some time in the staff spaces; the main staff room, the science base and the prep room and, if possible, take the opportunity to speak to one or two existing members of staff, teachers and support staff, including science technicians, and ask them about their experiences in the school.

Of course, once you arrive in a school, you may find that, through your reflections, the ethos and vision of the science department do not align with your beliefs and values about teaching. To what view of science, science teaching and science learning do you subscribe? Are you even aware of others' views about the vision and ethos of the department? If so, do others in the department share this view? Often within busy science departments, such conversations may not be taking place; science department meetings can become monopolised by pragmatics and logistics. You may wish to suggest adding a rolling agenda item in meetings in order to provide space for these discussions. Of course, there is always the chance that such discussions may result in you feeling at odds with colleagues within the department or school. In such circumstances, it may be useful to look at other departments, or even other schools, where you could work. It may be that a new school could provide a better 'fit' and the reinvigoration that you are looking for.

Professional development

Finally, reflections may suggest that a teacher desires or needs professional development (PD) to feel more empowered to successfully sustain his/her position. Traditionally, PD framed as continuing professional development (CPD) is one of the first ports of call when considering teacher retention. Indeed, teachers develop and hone their teaching skills throughout their careers and CPD ensures that teachers are engaged in the process of learning for their entire careers. We have consciously repositioned it further down the agenda; this is not because we do not believe CPD to be important, but because we see it framed within the context of what the individual needs are and how they are best met.

Adey (2011) refers to the 'drivers' of professional development as being *'conceptual change, reflection on practice* and *intuitive knowledge in teaching practice'*. These three strands of thought on the nature of PD are interconnected and feed into each other. This PD covers a range of different aspects, from formal CPD designed to develop subject knowledge, pedagogical skills, curriculum awareness; more informal professional development activities involving peer collaboration and engagement with fellow teachers; continuing with further studies such as completing a Masters or a PhD; or developing other professional skills such as management and leadership skills. Subject-specific CPD can be accessed through various sources:

- ➤ Learned Societies (e.g. Institute of Physics, Royal Society of Chemistry/Biology);
- ➤ Professional Associations and Networks (e.g. ASE, STEM Learning, STEMNET);
- ➤ School/Higher Education Institution (HEI)-provided; and
- ➤ A range of private providers, which can be accessed through your school.

While more general pedagogy-focused CPD can be found through HEI education departments, curriculum-targeted CPD might be found via examination boards or relevant government departments. As has been suggested throughout this chapter, conducting an audit of our own needs may help frame the type of CPD we are looking for and therefore the questions to ask of providers to ensure that we access the right CPD for our own needs. For example, what will be covered? Over what time period? What is the balance between input and sharing?

However, it is also useful to see CPD opportunities as spaces in which to reflect, in both formal and informal ways. Look for CPD that does not solely offer skills and curriculum development, but also personal development in the context of the role of a science teacher.

Having opportunities to develop professionally in their jobs is crucial for teachers. In many schools, there are opportunities to progress relatively swiftly up the career ladder. Indeed, as we accrue new roles and responsibilities in our career, we will need to learn new skills, learn to see new and different perspectives and develop new understandings. Therefore, seek out professional development opportunities to respond to the development of and changes in your career. By engaging in professional development, teachers can find a way to grow and change in their schools. Indeed, it may be that we *have* to change if we are to sustain our enthusiasm and energy for our work. For many teachers, there is the need to actively explore different avenues within their careers in order to maintain *'meaningful career trajectories'* (Rinke, 2014, p.66) in their schools.

In conclusion

In this chapter, we have considered a variety of ways in which someone might position him/herself as a science teacher after the ITE year. Career trajectories are neither homogenous nor linear; each science teacher travels his/her own personal journey through the profession. Here, we have attempted to highlight the various ways a teacher might seek out professional development experiences and enhance his/her science teaching career; this involves active reflection, engagement with research and further education/qualification. In this way, we are better able to navigate our teaching journey, make informed decisions and maintain the initial motivations that brought us into the science teaching profession.

Further reading

Huberman, A.M. *et al* (1993) *The Lives of Teachers*. London: Cassell

Knapp, M.S. & Plecki, M.L. (2001) 'Investing in the renewal of urban science teaching', *Journal of Research in Science Teaching*, **38,** (10), 1089–1100

Lynch, S., Worth, J., Bamford, S. & Wespieser, K. (2016) *Engaging Teachers: NFER Analysis of Teacher Retention*. Slough: NFER

Moscovici, H. (2008) 'Science Teacher Retention in Today's Urban Schools: A Study of Success and Failure', *Urban Education*

Olsen, B.S. (2010) *Teaching for success: Developing your teacher identity in today's classroom*. Boulder, Colorado: Paradigm Publishing

Worth, J., Bamford, S. & Durbin, B. (2015) *Should I Stay or Should I Go? NFER Analysis of Teachers Joining and Leaving the Profession*. Slough: NFER

References

Adey, P. (2011) 'What is next? CPD and the whole school'. In *Becoming a teacher. Issues in secondary education*, Maguire, M. & Dillon, J. (Eds.). Maidenhead: McGraw-Hill

American Psychological Association (APA) (2013) *How stress affects your health.* Available at: http://www.apa.org/helpcenter/stress.aspx Accessed 15.06.17

Briner, R. & Dewberry, C. (2007) *Staff wellbeing is key to school success*. London: Worklife Support Ltd/Hamilton House

Day, C. (2012) 'New lives of teachers', *Teacher Education Quarterly*, **39,** (1), 7–26

Day, C. & Gu, Q. (2010) *The New Lives of Teachers*. London: Routledge

Department of Education (DfE) (2016) *School workforce in England: November 2015*. (SFR 21/2016, 30 June 2016). London: DfE

Fokkens-Bruinsma, M. & Canrinus, E.T. (2012) 'The factors influencing teaching (FIT)-choice scale in a Dutch teacher education program', *Asia-Pacific Journal of Teacher Education*, **40,** (3), 249–269

Gu, Q. & Day, C. (2007) 'Teachers' resilience: A necessary condition for effectiveness', *Teaching and Teacher Education*, **23,** (8), 1302–1316

Gu, Q. & C. Day (2013) 'Challenges to teacher resilience: Conditions count', *British Educational Research Journal*, **39,** (1), 22–44

Heinz, M. (2013) 'Why choose teaching in the republic of Ireland? Student teachers' motivations and perceptions of teaching as a career and their evaluations of Irish second-level education', *European Journal of Educational Studies*, **5,** (1), 1–17

Hong, J.Y. (2012) 'Why do some beginning teachers leave the school, and others stay? Understanding teacher resilience through psychological lenses', *Teachers and Teaching*, **18,** (4), 417–440

Lave, J. & Wenger, E. (1998) *Communities of practice*. Retrieved from: http://valenciacollege.edu/faculty/development/tla/documents/Community ofPractice.pdf Accessed 16.06.17

Nias, J. (2002) *Primary teachers talking: A study of teaching as work*. London: Routledge

Manning, A. & Towers, E. (In press) 'Teachers' lives and careers: becoming and staying a teacher'. In *Becoming a Teacher*, Maguire, M. *et al* (Eds.). London: McGraw-Hill

Rinke, C.R. (2014) *Why Half of Teachers Leave the Classroom: Understanding Recruitment and Retention in Today's Schools*. Maryland: R&L Education

Roffey, S. (2012) 'Pupil wellbeing – teacher wellbeing: Two sides of the same coin?', *Educational and Child Psychology*, **29,** (4), 8

Troman, G. & Woods, P. (2000) 'Careers Under Stress: Teacher adaptations at a time of intensive reform', *Journal of Educational Change*, **1,** (3), 253–275

Watt, H.M., Richardson, P.W., Klusmann, U., Kunter, M., Beyer, B., Trautwein, U. & Baumert, J. (2012) 'Motivations for choosing teaching as a career: An international comparison using the FIT-Choice scale', *Teaching and Teacher Education*, **28,** (6), 791–805

Willis, J. (2014) 'Teacher's Guide to Sleep – and why it matters?', *The Guardian, November 11.* Retrieved from: www.theguardian.com/teacher-network/teacher-blog/2014/nov/11/good-night-teacher-guide-sleep

Worth, J. & De Lazzari, G. (2017) *Teacher Retention and Turnover Research. Research Update 1: Teacher Retention by Subject.* Slough: NFER

CHAPTER 22

Mentoring new science teachers

Michael Inglis

This chapter looks at the important role of mentoring student and newly qualified science teachers. It explains some of the approaches that can be taken and highlights particular issues that might be difficult to address. Using example case studies, the chapter encourages reflection on mentors' practice with a view to supporting mentors and improving the effectiveness of their work.

What this chapter is aiming to do

Mentoring new teachers is one of the most important and influential ways of contributing to the future of our profession. Mentoring is also a very effective way of developing our own practice as teachers. When supporting new teachers, most mentors discover that discussing and modelling their practice makes them aware of how much knowledge and wisdom about teaching they have accumulated, and stimulates them to reflect deeply on aspects of their practice that they have perhaps stopped thinking about.

This chapter will highlight some issues about mentoring new science teachers effectively. It is assumed that the reader is a teacher in a secondary school science department. The chapter is in two sections: the first provides a summary of information and expectations about mentoring new teachers, and the context of Initial Teacher Education (ITE). This is inevitably only a brief summary and sources of further information about mentoring are provided at the end of the chapter. The second section focuses on some key issues about mentoring new *science* teachers effectively.

A brief (perhaps contentious) point about terminology: a variety of terms is used for the process of becoming a secondary teacher, such as *initial teacher education, initial teacher training* and *pre-service education*. The various terms reflect different (often unconscious) beliefs about how people become effective teachers. I will use the term *initial teacher education* (ITE) to emphasise the idea that teachers should not simply be trained to perform a technical task (such as *what* to do and *how* to do it), but must also learn *why,* and how to develop their own practice through critical reflection. However, in places you will also see ITT (*initial teacher training*) referred to where reference is made to Department for Education (DfE) material.

Mentoring new teachers

Mentoring new teachers, whether they are student teachers on their first school-based placement or newly qualified teachers starting a new job, is rewarding and challenging. A good mentor does not just dispense information about improving practice. Effective mentoring involves establishing a supportive professional relationship, making appropriate judgements about when and how to challenge or intervene, and when to let a teacher experiment with his/her practice. How a mentor conducts the mentoring relationship, and how s/he models practice as a teacher and colleague, sends powerful messages to a new teacher about teaching as a *profession*.

In England, the *National Standards for School-Based Initial Teacher Training Mentors* (DfE, 2016a) define a mentor as:

'A suitably experienced teacher who has formal responsibility to work collaboratively within the ITT partnership to help ensure the trainee receives the highest-quality training' (p.7).

The National Standards are intended to specify the key skills and attributes of an effective mentor (see Figure 1 overleaf). Although the mentor standards are not statutory (and the emphasis on a training model is open to critique), they are expected to be used by ITE partnerships in England and *'Ofsted should have regard to the standards in their inspection of ITT providers'* (p.9). They are likely to inform mentor training for schools in England, and they provide a useful summary of what might be expected of a mentor in any national school context.

As we read these Standards, we should reflect on the attributes or skills that we think are already well developed in our practice, and the ones that we might need to develop further. Are there any standards that have particular implications or challenges for us as mentors of *science* teachers? How might we use the standards to inform our own professional development plan?

Having read the mentor standards overleaf, the reader should also review the other chapters in this book and reflect on how they relate to your practice as a mentor, as well as your practice as a teacher. Some universities provide Post-Graduate Certificate courses in mentoring, with the Open University providing

Figure 1: National Standards for School-Based Initial Teacher Training Mentors in England (DfE, 2016a pps.11–12)

Standard 1 - Personal qualities

Establish trusting relationships, modelling high standards of practice, and understand how to support a trainee through initial teacher training

The mentor should:

➤ be approachable, make time for the trainee, and prioritise meetings and discussions with them;

➤ use a range of effective interpersonal skills to respond to the needs of the trainee;

➤ offer support with integrity, honesty and respect;

➤ use appropriate challenge to encourage the trainee to reflect on their practice; and

➤ support the improvement of a trainee's teaching by modelling exemplary practice in planning, teaching and assessment.

Standard 2 – Teaching

Support trainees to develop their teaching practice in order to set high expectations of all pupils and to meet their needs

The mentor should:

➤ support the trainee in forming good relationships with pupils, and in developing effective behaviour and classroom management strategies;

➤ support the trainee in developing effective approaches to planning, teaching and assessment;

➤ support the trainee with marking and assessment of pupil work through moderation or double marking;

➤ give constructive, clear and timely feedback on lesson observations;

➤ broker opportunities to observe best practice;

➤ support the trainee in accessing expert subject and pedagogical knowledge;

➤ resolve in-school issues on the trainee's behalf where they lack the confidence or experience to do so themselves;

➤ enable and encourage the trainee to evaluate and improve their teaching; and

➤ enable the trainee to access, utilise and interpret robust educational research to inform their teaching.

Standard 3 – Professionalism

Set high expectations and induct the trainee to understand their role and responsibilities as a teacher

The mentor should:

- ➤ encourage the trainee to participate in the life of the school and understand its role within the wider community;
- ➤ support the trainee in developing the highest standards of professional and personal conduct;
- ➤ support the trainee in promoting equality and diversity;
- ➤ ensure the trainee understands and complies with relevant legislation, including that related to the safeguarding of children; and
- ➤ support the trainee to develop skills to manage time effectively.

Standard 4 – Self-development and working in partnership

Continue to develop their own professional knowledge, skills and understanding and invest time in developing a good working relationship within relevant ITT partnerships.

The mentor should:

- ➤ ensure consistency by working with other mentors and partners to moderate judgements; and
- ➤ continue to develop their own mentoring practice and subject and pedagogical expertise by accessing appropriate professional development and engaging with robust research.

a free online course through the Open Learn platform (see the information at the end of the chapter). There are also many books and online guides to generic mentoring, with White and Jarvis (2013) and Wright (2018) being two examples of books aimed specifically at the mentoring of teachers.

The ITE year

It might have been a long time since our student teacher days and there may have been many changes to the ITE system with which we will interact as a mentor. Information on current routes into teaching in each of the four UK nations can be found in the websites listed at the end of this chapter. As most readers of this chapter are likely to be in England, which currently has the most

Figure 2: QTS routes in England

PROVIDER LED		SCHOOL DIRECT		OTHER ROUTES
Led and managed by an accredited QTS provider in partnership with schools		Led by a consortium of schools which partner with an accredited QTS provider		
University-led Typically 60 days based in university Leads to QTS with PGCE (with M-level credits), or less common for secondary, BA/BSc with QTS	**SCITT (School-centred ITT)** Based entirely in, and run by, a consortium of schools A SCITT might also partner with a university to provide a PostGraduate Certificate qualification	**School Direct Fee** School consortium leads recruitment onto the QTS course and provides the placements Might also lead to PGCE, if School Direct consortium partners with an accredited University QTS provider	**School Direct Salaried** Student teachers are employed by a school, receive a salary and have their own classes. More commonly chosen by career changers or people already working in a school	**Teach First** **Troops to Teachers** **Researchers in Schools** **QTS Assessment-only** **Future Teaching Scholars**

complex system of ITE routes, the author will dwell a little more on this system. Figure 2 attempts to capture the diversity of English ITE routes that lead to QTS (Qualified Teacher Status), with some key features. In 2016/17, approximately 55% of new teachers went through a provider-led route (DfE, 2016c) and 45% chose to go through *School Direct* and other routes such as *Teach First*, but be aware that these proportions may well change significantly in coming years. QTS can only be awarded by an accredited QTS provider. Some ITE routes are led by an accredited provider (such as a university or a SCITT (School-Centred Initial Teacher Training)), and some are led by a group of schools that then selects an accredited provider with which to work (*School Direct*). All routes involve student teachers spending at least 120 days placed in two or more schools, where they will work with host teachers to take over the teaching of a variety of classes.

The accredited QTS provider is responsible for training mentors in the appropriate processes and procedures. For most mentors of student teachers, the main tasks are likely to include those shown in Figure 3.

Figure 3 is not intended to be an exhaustive list. For example, mentors should do some joint planning with their student teachers and encourage host teachers to do the same. Mentors should be aware of the ITE course

Figure 3: Most common tasks for mentors supporting student teachers

Creating a timetable

This is likely to start at around 30% of a full-time teacher's timetable during the early part of the 1st placement and will increase to 70% or more during the 2nd placement. The timetable should cover a range of science classes across the age range in your school. It should include where possible teaching of biology, chemistry and physics at Key Stage 3 (aged 11-14) so that the student teacher gains experience of teaching outside of their subject specialism. The timetable should also include solo teaching of post-16 classes if the ITE course requires it. It is good practice for a student teacher to be assigned to at least one of the mentor's own classes. If possible, the student should also be assigned to classes belonging to a range of science department colleagues.

Holding a regular individual meeting to review progress

This should be scheduled to allow time to discuss progress across all of the student teacher's classes, identify appropriate actions and agree developmental targets. The meeting should be in a quiet place that ensures privacy and where you are unlikely to be interrupted. The mentor and the student teacher should both come prepared for this meeting. The mentor should have received feedback from host-teacher colleagues about the student's teaching of their classes and the student teacher should be expected to have reviewed their progress since the previous meeting and be ready to discuss this. Discussion of the student teacher's subject knowledge and subject-specific pedagogy should feature prominently in these meetings.

Liaising with colleagues who are host teachers

Part of the mentor's role is to support colleagues who are acting as host teachers with conducting lesson observations, evaluating progress and dealing with problems. The mentor should collect feedback regularly from host teachers about the student teacher's progress with their classes. The mentor should also monitor the effectiveness of the student teacher - host teacher relationship, intervening and providing support where necessary.

Observing lessons

The ITE course will probably require a minimum number of formal observations per week and might provide a pro forma document on which the observer should record comments and targets. Prior to an observation, student teachers should be encouraged to identify aspects of their practice the observer should focus on, and communicate this information to the observer. The post-lesson meeting should be a dialogue that encourages the student teacher to reflect critically on their practice and identify how to develop it.

Assessing and reporting progress

Although the Teachers' Standards apply to all teachers in England (see Dept. for Education, 2011a), mentors should apply the Standards "to a level that is consistent with what should reasonably be expected of that teacher, given their role and level of experience" (Dept. for Education, 2011b). The Standards (or equivalent criteria used in your national context) should be used to frame discussions during regular review meetings and setting of targets. The ITE course will have a system in place for student teachers to record evidence of progress. The mentor might be required to review and formally approve the evidence collected by the student. The mentor may also be required to formally assess and report a student teacher's progress at fixed points in the placement.

procedures for working with a student teacher whose rate of progress, or professional conduct, is causing concern. There may also be specific mentoring tasks associated with particular ITE routes.

The post-ITE induction year

Once teachers have passed their ITE course in England and Wales, they must then complete an induction year as a Newly Qualified Teacher (NQT) and, in Scotland, as a probationer teacher. It is important to remember that Initial Teacher Education or Training is exactly that, *initial*. Teachers will continue to develop their practice throughout their careers, and the induction year is a critical stage for developing confidence and experience. The requirements for what a school must provide during the induction year are specified for each UK nation (see the websites at the end of this chapter). In England, these requirements include the need for an NQT to have a timetable of no more than 90% of a main scale teacher in that school and entitlement to a planned induction programme of supportive observations and review meetings with a designated mentor. You can find additional guidance about the induction year from the various teaching unions (e.g. NEU, NASUWT, EIS)[1]. In a role as an induction mentor, the starting point when developing an induction programme should be the strengths and areas for development identified at the end of the new teacher's ITE course.

Mentoring new *science* teachers

Many of the skills of science teaching can be difficult to articulate to others, because teaching is such a complex activity, requiring rapid professional judgements about actions to take during real time evaluation of the often-competing demands of a group of children.

As the mentor standards above indicate, along with providing advice, a mentor will need to work with new science teachers so that they can reflect on their problems, make good judgements about their practice and develop positive relationships with children. Along with these requirements, mentoring new *science* teachers presents some particular challenges compared to mentoring in many other subjects, with some of the most common being:

> ➤ Science departments tend to be large. It can be easy for an NQT or a student teacher to feel overwhelmed by negotiating a large web of relationships with new colleagues, while also keeping on top of the demands of being a new teacher;

[1] NEU is the National Education Union, formed from the merger of the NUT and ATL

➢ The prominence afforded to science in the school curriculum can lead to additional stress for new teachers, and for mentors trying to balance the needs of new teachers with the demands on the science department's examination results;

➢ Most science teachers teach, for at least some of the time, outside their science subject specialism: for example, in England usually at Key Stage 3 (age 11-14), but often also at Key Stage 4 (age 14-16). New teachers need to develop their own deep understanding of science concepts that they might not have studied since they themselves were at school. They also need to develop their pedagogical content knowledge to know children's likely conceptual difficulties and how best to address them. Teaching outside their subject specialism can be an intimidating prospect for many teachers, but especially for new teachers, who are unable to draw on experience when trying to design lessons that will engage children; and

➢ Practical work is seen by children as a distinctive feature of school science, but using practical work *effectively* to enhance children's learning can be problematic (see Chapter 12).

Some of these issues will be illustrated and explored through the following short descriptions of new teachers, Nigel, Mary and Tigist. The cases have been constructed from experience of working with many new teachers. The comments that follow the cases present possible interpretations and implications for good mentor practice. The intention is to highlight some of the more common issues faced by science mentors. We will have our own additional interpretations and thoughts about actions as a mentor and, hopefully, it will be useful to reflect on how our interpretations relate to the ones shown here. As we read Cases 1, 2 and 3, we should consider the issues that they raise for us as science mentors. Which of these issues do we think are the most important to address? Why do we think this? How could we plan to work with these new teachers to improve their practice?

➢ **Case 1:** Nigel is a student teacher who has just started his second school-based placement. Teachers who have observed Nigel's lessons report that he has good subject knowledge, a positive and supportive manner with many of the students, and tries hard to plan engaging and well-structured lessons. However, he sometimes pitches the level of his explanations of science concepts too high. His body language and use of voice can communicate nervousness and lack of confidence. During lessons, Nigel has a tendency to ignore certain students who are chatting and off-task, and those students appear increasingly unconcerned about Nigel's presence in the room. During one of his regular meetings with his mentor, Nigel reveals that he is uncomfortable with confronting challenging behaviour and dealing with children who might be rude to him, and feels embarrassed about admitting this.

Comments on Case 1: Nigel

The immediate causes of the problems that Nigel is experiencing might be obvious to some. Clearly, Nigel needs to be aware of the messages he communicates through his body language and the necessity to deal with challenging behaviour assertively. He also needs to recalibrate his judgements about the suitability of his explanations for the students in his class. However, simply telling this to Nigel is unlikely to address his problems effectively. Mentoring Nigel skilfully may involve encouraging him to reflect on how he feels when students ignore him, how the students might feel in his lessons when they cannot follow his explanations, and what he understands about behaving assertively. The mentor will need to be sensitive in drawing Nigel's attention to his body language and perhaps rehearse with him how he can move around the room in a way that portrays confidence. This might include the mentor sharing his/her own experiences of improving self-awareness and learning how to 'be' a teacher who has presence, despite feelings of nervousness. In addition, the mentor could arrange for Nigel to observe experienced colleagues dealing with off-task behaviour who will also talk with him about how they modify their own body language and the complexity of their explanations to keep students engaged. The mentor could ask host teachers who are observing Nigel's lessons to focus on feeding back to him about his body language and how successfully he adapts his scientific explanations as he experiments with his practice. These kinds of actions by the mentor are likely to not only help Nigel improve his practice, but will model to him that effective and experienced teachers also experience challenging behaviour from children and work collegially to support each other.

➤ Case 2: Mary is a biology graduate who is making good progress with teaching her mentor's biology Key Stage 4 classes (the mentor is a biology specialist). An ITE course tutor visits to co-observe, with the mentor, Mary teaching the topic of forces to a Year 8 class (age 12-13). During the lesson, the mentor identifies areas of strength and potential development for Mary's use of assessment and her organisation of activities. The course tutor joins the post-lesson discussion with Mary. Mary is reflective and engages in a critical and fruitful discussion with the mentor, covering unprompted all the key points the mentor had identified about those two aspects of Mary's practice. The mentor agrees appropriate development targets and is happy with the outcome and conduct of the discussion. The ITE tutor then asks Mary about some of the students' forces misconceptions that were revealed by one of the activities, and how she might design her next lesson with those misconceptions in mind. Mary is surprised by this line of questioning.

Comments on Case 2: Mary

This is a common situation for student teachers: teaching outside their subject specialism. The fact that Mary is 'making good progress' suggests that the mentoring has perhaps been effective in supporting Mary to form productive relationships with her classes and teach biology so that the children make progress. Although the mentor identified valid areas for development (use of assessment and organisation of activities), they, and the subsequent targets, are not science-specific. Mary's surprise at the ITE tutor's questions about children's misconceptions is perhaps indicative that such discussion is not a regular feature of the mentoring she experiences. It is also possible that both the mentor and Mary, as biology specialists, might not have been aware of the forces misconceptions that the ITE tutor noticed. A common criticism by Ofsted of mentoring in schools in England is that much of the target-setting seen across all school subjects is not specific to children's learning of that subject, and tends to focus on generic aspects of practice, such as classroom management. Research has shown that, when science teachers plan lessons outside their science specialism, or have insecure understanding of their own subject, they tend to draw predominantly on information from textbooks and websites, and they are likely to over-rely on a pre-prepared presentation of information. When teaching within specialism, teachers are more likely to use strategies that require children to participate actively in lessons (Kind, 2009; Hattie, 2009; see also Chapter 7). It is important that mentors place subject knowledge, and how it can be used to inform effective planning, at the centre of the regular mentor review meetings and during post-lesson observation discussions. This means that mentors need to be secure in their subject understanding or be able to seek support from colleagues where needed.

➤ Case 3: Tigist is a chemistry specialist who is teaching distillation to a Year 9 class (aged 13-14). She has planned for the class to do practical work involving distillation of brine. In her lesson plan, the learning objectives refer to revising mixtures and compounds, explaining changes of state in terms of particles and being able to explain how to use distillation to separate mixtures. After reviewing what the students remember about mixtures and compounds, Tigist demonstrates how to assemble the equipment and then instructs the class to carry out the practical. During the practical, the NQT mentor observes several safety issues and instances of students struggling to carry out the practical task correctly. Tigist works hard to circulate during the practical and support students to address many of these issues. All the student talk that the mentor overhears is about the equipment and following the instructions correctly. Not all students complete the practical, despite Tigist allowing the task to overrun her planned amount of time. During the somewhat rushed plenary, some of the students' responses reveal that they still have misconceptions about distillation and what is happening during a change of state.

Comments on Case 3: Tigist

For many science teachers, practical work is such an intrinsic aspect of their practice that it can be easy to stop thinking about its purpose and whether it is effective. As discussed in Chapter 12, the notion of *effectiveness* in practical work requires careful thought, if a teacher's intention is to use it to help students to develop conceptual understanding. Tigist's lesson is fairly common for NQTs and student teachers. For this class, the stated learning objectives are perhaps appropriate, but it appears that Tigist has not considered properly how the practical work will address them. The fact that the students' talk is all about the procedure and the equipment, and not about the underlying science, is perhaps a consequence of Tigist not explaining the purpose of the practical task in terms of the lesson objectives. If the practical is intended primarily to help the students to develop equipment manipulation skills, then this needs to be made clear to them. If the practical is really intended to focus on conceptual understanding of the particle model, then a teacher-led demonstration with good use of questioning might be more effective. Given the difficulty that Tigist has experienced in getting the students to complete the practical in the allotted time, a demonstration may well have been the better choice to make.

So what can effective science-specific mentoring look like? As we read the example of Wilma's mentoring of Lee (Case 4), we should reflect on how her approach relates to our own experiences of being mentored. What might be effective about Wilma's mentoring? What else could she do?

➢ **Case 4:** Wilma is mentoring Lee, a student science teacher. Over the course of several lessons, Lee has been consistently struggling to differentiate effectively his teaching of electric circuits so that all his Year 8 students are making progress. Wilma addresses this by teaching the next lesson herself in order to model some approaches to differentiation, while Lee observes the lesson. During the post-lesson meeting with Lee, Wilma models how she reflects on her own teaching by talking through why she made certain pedagogical choices to address specific student conceptual difficulties about parallel circuits, and encouraging Lee to discuss what he noticed about their effects on the students' learning. Wilma then moves the discussion on to Lee's knowledge of students' likely misconceptions about electric circuits, and how he might be able to use this knowledge to adapt his practice with all his classes. Once Lee has identified some specific actions that he could try out, he and Wilma then plan collaboratively Lee's next Year 8 lesson about voltage, with Wilma encouraging Lee to justify his choices of activities. Wilma then prompts Lee to identify three targets about developing his use of differentiation with all his classes in the coming week, including a target about selecting a particular differentiation approach that might be suitable for teaching voltage to his Year 11 class (aged 15-16).

Comments on Case 4: Wilma and Lee
Rather than simply advising Lee on differentiation approaches to use, Wilma chooses to model her own practice. Notice that she makes the reasons for her choices of pedagogy explicit during the post-lesson discussion. It is this explicit modelling that can be such an effective developmental tool (Loughran & Berry, 2005; Lunenberg et al, 2006), and is reflected in Mentor Standard 1 (Figure 1). It can be effective because Lee sees practice in action, he hears about and discusses the thinking process that underpins it and, crucially, he is also experiencing directly how a teacher might reflect on his/her own practice. Wilma anchors the discussion of her practice in terms of the conceptual knowledge that the students are supposed to be learning, and she seems to expect Lee to do the same. This is important, because student teachers often need to make a transition from an understandable focus on executing teaching activities as planned, to a focus on the effect of those activities on students' learning. By planning collaboratively with Lee, Wilma can continue to model her practice explicitly and encourage Lee to articulate how his teaching will focus on children's learning. Wilma appears to expect Lee to take responsibility for developing his practice, by prompting him to identify subject-specific targets rather than simply telling him what his targets will be. By expecting Lee to take

ownership of developing his practice as a student teacher, he might be better prepared to take control of his development as an NQT and beyond. Expecting at least one of the targets to be science-specific could help Lee avoid the common new-teacher pitfall of focusing only on general teaching activities or classroom management.

Some final points to consider when mentoring new science teachers

First of all, when mentoring a student teacher, find out about available mentor training and support from the ITE course provider, especially science-specific mentor training. Mentors are expected to develop their practice (see Mentor Standard 4 in Figure 1) and should be supported to do so. Mentors should make sure that they are familiar with the reporting and assessment requirements for the ITE course, or induction year, so that they can focus their time and energy on working with the new teachers instead of negotiating procedures.

Some attention to general pedagogical skills and classroom management will be necessary, especially in the early stages of a student teacher's placement or as a newly qualified teacher becomes established. However, a mentor's role is about supporting new teachers to become excellent *science* teaching practitioners, and this requires a focus on science subject knowledge and using it to inform pedagogy. Notice that Mentor Standard 2 (Figure 1) states that a mentor should *'support the trainee in accessing expert subject and pedagogical knowledge'*. It is not feasible for most teachers to develop a deep understanding of all aspects of school science subject knowledge within the time constraints of an ITE course. Research shows that student teachers tend to develop their subject knowledge as and when it is needed to teach an upcoming topic, and that they want the support of school colleagues as sources of topic-specific pedagogy (Lock *et al*, 2011). A mentor is not expected to know everything there is to know about expert science teaching. However, s/he will be expected to know how to support a new teacher to access this knowledge (see Mentor Standard 2 in Figure 1). Science colleagues (and other teachers in the school) constitute a pool of knowledge and experience from which mentors can draw.

The Association for Science Education (ASE) provides a range of publications and resources to support science teachers, especially the journal *School Science Review*, which is aimed at science educators and shares good practice, expert advice and relevant research in science education. Mentors should be aware of professional bodies such as the Institute of Physics, the Royal Society of Biology and the Royal Society of Chemistry, which all provide resources and support for teachers (see, for example, the *Supporting Physics Teaching* materials on the Institute of Physics website).

Managing practical work effectively is a commonly cited concern for new teachers. During an ITE course, it is not possible for student teachers to acquire all the practical experience they need to develop confidence. New teachers should be encouraged to talk with the science technicians and to gain experience with practical equipment and procedures prior to using them in lessons. The mentor and host teachers should prompt student teachers to think critically about their use of practical work, so that they plan to use it effectively (see Chapter 12).

Encourage host teachers to do some joint lesson planning and team-teaching with their student teachers, especially when they are teaching outside their subject specialism. This can help the student teachers to learn how to make lesson planning manageable while focusing on making lessons effective.

Much of what a new teacher learns about being an effective science teacher comes from participating in the life of the science department and working alongside colleagues. This includes being included in the informal conversations over coffee at break times. Student teachers should be expected to attend, and perhaps contribute to, science department meetings. All colleagues in the department should be encouraged to share resources and talk about their practice.

Finally, new teachers bring new ideas and practices, as well as energy and enthusiasm. They should be encouraged and supported to be innovative. We all have much to gain from working with them.

Further reading

Information on the ITE and induction year systems for the UK nations:

Department for Education (2016b) *Get into Teaching*. Available at: https://getintoteaching.education.gov.uk Accessed 30.06.17

Department for Education (2015) *Carter review of initial teacher training.* Available at: https://www.gov.uk/government/publications/carter-review-of-initial-teacher-training Accessed 01.09.17

Department for Education (2013) *Induction for newly qualified teachers (NQTs).* Available at: https://www.gov.uk/government/publications/induction-for-newly-qualified-teachers-nqts Accessed 30.06.17

National Union of Teachers (2017) *NQT induction guide.* Available at: https://www.teachers.org.uk/sites/default/files2014/induction-guide-20pp-2017-11046.pdf Accessed 09.09.17

General Teaching Council for Scotland (2016) *Teacher Journey*. Available at: http://www.gtcs.org.uk/TeacherJourney/teacher-journey.aspx Accessed 01.09.17

Teacher Training and Education in Wales (2017) *Teacher training and education in Wales*. Available at: http://educationcymru.org/home Accessed 01.09.17

Department of Education Northern Ireland (2017) *Initial teacher education courses in Northern Ireland*. Available at: https://www.education-ni.gov.uk/articles/initial-teacher-education-courses-northern-ireland Accessed 01.09.17

Mentoring:

ATL (2012) *Guide to Mentoring*. Available at: https://www.atl.org.uk/advice-and-resources/publications/guide-mentoring Accessed 01.09.17

Department for Education (2016a) *National Standards for School-based Initial Teacher Training (ITT) mentors*. Available at: https://www.gov.uk/government/uploads/system/uploads/attachment_data/file/536891/Mentor_standards_report_Final.pdf Accessed 30.06.17

OpenLearn (2017) *Learning to teach: mentoring and tutoring student teachers*. Available at: http://www.open.edu/openlearn/education/learning-teach-mentoring-and-tutoring-student-teachers/content-section-0 Accessed 01.09.17

White, E. & Jarvis, J. (2013) *School-based teacher training: a handbook for tutors and mentors*. London: Sage

Wright, T. (2018) *How to be a brilliant mentor. 2nd Edition*. Abingdon: Routledge

Other sources of useful information:

Association for Science Education (2017) Available at: https://www.ase.org.uk/home Accessed 09.09.17

Institute of Physics (n.d.) Available at: http://www.iop.org/education/index.html Accessed 01.09.17

Institute of Physics (n.d.) *Supporting physics teaching*. Available at: http://www.supportingphysicsteaching.net Accessed 08.07.17

Royal Society of Biology (n.d.) Available at: https://www.rsb.org.uk/education Accessed 01.09.17

Royal Society of Chemistry (n.d.) Available at: http://www.rsc.org/resources-tools/education-resources Accessed 01.09.17

References

Department for Education (2016c) *Initial teacher training: trainee number census – 2016 to 2017*. Available at: https://www.gov.uk/government/statistics/initial-teacher-training-trainee-number-census-2016-to-2017 Accessed 08.09.17

Department for Education (2011a) *Teachers' Standards*. Available at: https://www.gov.uk/government/publications/teachers-standards Accessed 30.06.17

Department for Education (2011b) *How you should use the teachers' standards*. Available at: https://www.gov.uk/government/publications/teachers-standards Accessed 30.06.17

Hattie, J. (2009) *Visible Learning*. Abingdon: Routledge

Kind, V. (2009) 'Pedagogical content knowledge in science education: perspectives and potential for progress', *Studies in Science Education*, **45,** (2), 169–204

Lock, R., Salt, D. & Soares, A. (2011) *Acquisition of Science Subject Knowledge and Pedagogy in Initial Teacher Training* [Online]. London: The Wellcome Trust. Available at: www.wellcome.ac.uk/About-us/Publications/Reports/Education/WTVM053203.htm Accessed 05.03.14

Loughran, J. & Berry, A. (2005) 'Modelling by teacher educators', *Teaching and Teacher Education*, **21,** 193–203

Lunenberg, M., Korthagen, F. & Swennen, A. (2006) 'The teacher educator as a role model', *Teaching and Teacher Education*, **23,** 586–601

CHAPTER 23

Leading science education in a secondary school science department

Helen Gourlay & Euan Douglas

This chapter looks at leading a science department and therefore science education in a school. This task is a complex one, which often involves leading a large and diverse team of science teachers and technicians, as well as managing specialist facilities, resources and equipment. The chapter starts by looking at what school effectiveness research identifies as good practice in leading and managing a science department and expands on these key areas to support heads of science in effective leadership.

What does school effectiveness research say about departments?

Whilst national league tables continue to rank examination outcomes at the whole school level, school effectiveness research suggests that there may be significant differences in effectiveness between different subject departments in the same school. An example of research at the departmental level is Sammons et al (1995) who studied departmental effects at GCSE. They compared ninety-four schools' results in English, English literature, mathematics, French, history and science, with the outcomes predicted based on school background factors – students' age, gender, ethnicity and eligibility for free school meals. They analysed schools' examination results over three years to find two schools that were more effective than others, two that were less effective, and two that had a mixture of more effective and less effective departments. The Headteachers, Deputy Headteachers and Heads of Department (HoDs) in these six schools were then interviewed to identify characteristics that might make them differentially effective.

These findings were then tested using a questionnaire, which was completed by the Heads, Deputy Heads and HoDs in ninety schools. Their responses were compared with the analysis of their schools' effectiveness as evaluated in the earlier study.

Some of the characteristic activities of good HoDs were introducing new ideas, carrying out curriculum planning, and teaching by example (modelling good teaching). Where HoDs felt that their departments had a clear sense of direction, the value-added GCSE scores were better. Unity within the department also had a positive effect: value-added results were better where HoDs felt that there was a consensus about educational philosophy and goals within the department, and where there was involvement in decision-making about whole school policies. Effects were negative where there had been disagreement on student groupings, or lack of cohesiveness in the department. Where HoDs reported a consistency of approach in their departments, such as a consistently applied homework policy or marking policy, value-added results were better.

Because they have been deemed to be indicative of effective departments, this chapter will consider the following aspects of the role of HoD:

- ➢ Developing a consensus about philosophy and goals;
- ➢ Introducing new ideas;
- ➢ Curriculum planning and modelling good teaching;
- ➢ Developing a clear sense of direction: Development planning; and
- ➢ Developing consistency through implementation of departmental policies.

Additionally, we have chosen to focus on retention of teachers – this is because schools are facing challenges in this area currently. Difficulties in retention are believed to be particularly acute in science (Allen & Sims, 2017). The improvement in science teachers' teaching is rapid in their first few years in the profession, and high staff turnover is thought, therefore, to be damaging to students' learning.

Developing consensus about philosophy and goals

In a previous edition of this *ASE Guide,* Oswald (2006, p.186) stated that 'the *role of a subject leader is to secure good progress and attainment in science for young people'*. This view is consistent with those of school effectiveness researchers, such as Scheerens (1992, p.79), who stated that '*effectiveness is apparent from education results'*. He justified the emphasis on examination results, rather than on other educational objectives, such as the ability of students to co-operate, or the development of students' independence, in three ways (p.88):

- ➢ If these aspects of education are not examined, they cannot be deemed by anyone to be very important.

- They may not be teachable, or measurable.
- They may contribute to students' learning and thus have an indirect effect on students' achievements in examinations.

Some would argue that this is a rather narrow view. At the time of writing this chapter, the new head of Ofsted, HMCI Amanda Spielman, has given a very welcome speech, in which she said:

'One of the areas that I think we sometimes lose sight of is the real substance of education. Not the exam grades or the progress scores, important though they are, but instead the real meat of what is taught in our schools… Yes, education does have to prepare young people to succeed in life and make their contribution to the labour market. But to reduce education down to this kind of functionalist level is rather wretched.

'Because education should be about broadening minds, enriching communities and advancing civilisation. Ultimately, it is about leaving the world a better place than we found it' (Spielman, 2017).

Whilst we acknowledge the constraints imposed by national curricula, examination specifications, external assessments, inspections and other accountability measures, the speech suggests that we may be experiencing a slight change of climate. There is perhaps some scope for us to involve our science teams in developing a collective vision for science learning in our schools, going beyond aiming to achieve good examination results (as important as that is).

Two of the key debates about the science curriculum in England have centred around the following questions:

- To what extent should we be teaching science as a body of factual knowledge, as opposed to giving students a good understanding of the nature of science?
- To what extent should we be preparing students to be future scientists, as opposed to preparing all students for citizenship?

Whilst supporting the idea that we should be developing the aspirations of all students, the answer to these questions may be different in different schools. The emphasis we place on them in our departments needs discussion because it will affect teaching approaches.

Introducing new ideas: Case study – Implementing an assessment system for life without levels

Saint George Catholic College, Southampton (Douglas, 2017), written by Euan Douglas:

"Nick Gibb (Minister of State for Schools) stated in a speech that National Curriculum levels were originally introduced in England *'with the intention of delivering an assessment system which measured students' progress against a national framework'* (Gibb, 2015). However, when they were removed in 2015, no system was introduced in their place; instead, individual schools were given the opportunity to devise their own systems.

At Key Stage 3 (age 11-14), we refocused on encouraging students to work scientifically and to use scientific ideas to explain things. We believed that concentrating on these two strands across Key Stage 3 would be the best foundations for students when they reached GCSE.

Removing levels meant that our team needed our own progression framework that was consistent across the department. This was important because;

➢   students move groups and teachers throughout their time in school, so a consistent approach reduces the variation students experience; and

➢   our whole school marking policy expects frequent written feedback and, as less experienced teachers may find it hard to give suitable feedback, a consistent approach helps them identify which areas to prioritise.

Therefore, there needed to be a balance between allowing teachers freedom in and responsibility for assessing their own classes, whilst ensuring consistency across the department. I found that the best way to achieve this was to introduce assessments that all classes took, whilst allowing teachers freedom to plan individual lessons and sequences of lessons in their own way (supported by a department scheme of work and resource bank to reduce workload).

When I began designing these assessments, my priority was to develop a method of assessing the same Assessment Objectives as at GCSE, but in a way that was appropriate for younger and less experienced students. Along with my Second in Department, we designed a series of consistent and high quality assessments that were effective measures against our key priorities of working scientifically and giving explanations. Our approach was based on the SOLO taxonomy (Biggs & Collis, 1982). With our priorities and assessments shared with the department, they informed and guided everyone's teaching, which aided the drive for consistency in outcome rather than teaching approach.

☞

Any change in practice like this had the potential for colleagues deciding not to change but continue doing things their own way. I avoided this by removing the obstacles. Our work ensured that they no longer spent significant periods of time preparing and designing assessments. Instead, this time could be better spent preparing high quality lessons and giving good quality, subject-specific feedback. I ensured that the department was involved in the process by asking for feedback at different stages. It was important to show how colleagues' inputs and concerns had been acted upon, so that it was a genuine team development.

Once the new framework was in place, it was important to make sure that colleagues were following it. This was to ensure that support could be offered where needed, and that priorities and goals had been communicated clearly. This was done through book audits and devoting time in department meetings to discuss its implementation.

Our department was going through this process at the same time as the rest of the school. By being proactive, and meeting the deadlines set by the Senior Leadership Team (SLT), we were successful in developing a framework that worked well for science."

Curriculum planning and modelling good teaching

In terms of planning the science curriculum and leading teaching, there are several ideas that science team leaders may wish to consider. We recognise that this is not straightforward, owing to the many constraints within which we work (not limited to national curricula and examination specifications). However, we would encourage readers to do so, as they may have the potential to support you in continuously improving your students' science education.

Arguably, one of the main benefits of the introduction of a national curriculum in England was greater uniformity of students' experiences of science in primary schools (Millar & Osborne, 1998). According to Ofsted (2015), Key Stage 3 in England remains an area for development, although this probably varies markedly within and between schools. Whilst acknowledging that it is difficult to do in practice (not least because secondary schools may have numerous different primary feeder schools), we recommend that secondary science leaders consider how best to build on students' earlier experiences of science education.

A number of issues that may be of interest to science leaders when planning the curriculum arise from science education research.

Alternative conceptions research (see, for example, Driver *et al*, 2014) tells us that children arrive in our science lessons with preconceptions about science phenomena. Their naïve ideas may be at odds with the accepted scientific view and may be hard to shift. Whilst there is not a consensus about how best to overcome these difficulties, research suggests that outcomes are better where teachers are aware of children's possible alternative ideas (diSessa, 2005). Departmental planning documents could make common misunderstandings and misconceptions clear to teachers. See, for example, the Piper project monographs about teaching electricity and light (Institute of Physics, 2017).

Readers may also wish to consider what we can learn from the original CASE project (Adey & Shayer, 2006). This was a teaching intervention in the lower secondary age range, which produced outcomes suggesting that cognitive gains could be made. The mechanism by which the CASE project produced gains was not clear, although it may have been related to the emphasis on student talk and discussion, and on metacognitive strategies. Unfortunately, a recent trial by the EEF showed no gains, but this was thought to be related to difficulties in implementation (EEF, 2017).

When considering the curriculum offer for Key Stage 4 in England (Years 10 and 11, students aged 14-16), science subject leaders should be aware of the emerging debate about double science versus triple science. The ASPIRE team's research (Archer *et al*, 2016) suggests that students' perceptions of whether or not they should study science post-16 are affected by the options available to them at age 14-16. Those doing double science may get the impression that they are not clever enough to study the sciences post-16, when this may not be the case. The team also raised a concern that guidance given at age 14 may be disadvantageous towards students from low socio-economic status backgrounds.

It has also been suggested by a group of international scientists, engineers and science educators that students' science learning would benefit from being organised around big ideas (Harlen, 2010, 2015, see Figure 1 overleaf). The rationale of organising science ideas in this way is to help children make sense of school science, and to understand its relevance to the world around them.

At a more practical level, we suggest that there are three levels of curriculum planning that should be carried out:

➢ Long-term plans, indicating which topics are to be taught in which year, by whom and in what order. Staggering topics through the year for different student groups may impact on moving students between groups, but will enable clashes to be avoided for limited supplies of equipment, as well as taking into account seasonal variation (for example, pond-dipping does not go so well in the winter months, as pond life may be hibernating (or dead), and testing leaves for starch will go better when your geraniums are flourishing!);

➤ Medium-term plans or schemes of work, which suggest teaching sequences within topics, and refer to specific resources and assessment opportunities; and

➤ Detailed lesson plans, which may be accompanied by lists of practical and other teaching resources, worksheets, PowerPoint presentations, video clips, etc.

Figure 1: Fourteen big ideas in science (Harlen, W. (Ed.), 2010)

Ideas of science

1. All material in the Universe is made of very small particles.
2. Objects can affect other objects at a distance.
3. Changing the movement of an object requires a net force to be acting on it.
4. The total amount of energy in the Universe is always the same, but energy can be transformed when things change or are made to happen.
5. The composition of the Earth and its atmosphere and the processes occurring within them shape the Earth's surface and its climate.
6. The solar system is a very small part of one of millions of galaxies in the Universe.
7. Organisms are organised on a cellular basis.
8. Genetic information is passed down from one generation of organisms to another.
9. Organisms require a supply of energy and materials for which they are often dependent on or in competition with other organisms.
10. The diversity of organisms, living and extinct, is the result of evolution.

Ideas about science

11. Science assumes that for every effect there is one or more causes.
12. Scientific explanations, theories and models are those that best fit the facts known at a particular time.
13. The knowledge produced by science is used in some technologies to create products to serve human ends.
14. Applications of science often have ethical, social, economic and political implications.

There is debate amongst colleagues about the extent to which it is helpful to have detailed departmental lesson plans. On the one hand, they can significantly reduce the amount of time that individual teachers spend on planning, particularly for less experienced colleagues such as newly qualified teachers. On the other hand, where departmental plans are highly prescriptive, they may limit colleagues' ability to learn to plan lessons for themselves, and make it difficult for them to develop their own teaching styles. Teachers will need to adapt lessons somewhat to meet the needs of specific groups of students, however detailed the departmental plans.

Our suggestion would be to adopt approaches that reduce colleagues' workloads as much as possible, in light of current concerns about teacher retention. A possible model would be to delegate planning for particular topics to different members of the team and to share resources. This is particularly useful in departments with a high rate of staff turnover, as it enables colleagues joining the department to pick up where others have left off, which should help to ensure continuity of learning experiences for the children. Collaborative planning or projects in which groups of teachers research their own teaching may be even better, but can be difficult to schedule given teachers' workloads. One approach that may be fruitful is Japanese lesson study (Dudley, 2011).

Developing a clear sense of direction: Development planning

The school effectiveness research highlighted the importance of having a clear sense of direction and shared goals within the science team. In most schools, HoDs produce a departmental development plan and share this with both their line manager and their team. It is best not to see each year in isolation, but as part of a longer plan for development.

A natural point to review the effectiveness of a department is following the results of external examinations in August, and this also allows us to start a new academic year communicating the plan to the team, including any new staff. Look at the areas you were developing last year, and evaluate if progress has been made in line with what you planned or not, before deciding whether they would remain one of your development areas for the coming year. Not allowing too many areas or targets to appear on the development plan is important, so that all colleagues can be clear on their importance; try to aim to have three or four targets.

To increase the buy-in from members of your own department and also to garner support from the SLT, try to link your own department's development plan to the school's plan. Ideally, use your department plan to show how you will implement the school's priorities within your team.

One other strategy that can help the science team engage with a development plan is to link colleagues' performance management targets to aspects of it. This offers science teachers an opportunity to lead aspects and gives them responsibility, as well as providing a formalised way of holding them to account.

Once the development plan is written, shared and communicated, go back to it regularly during the year. Use it to keep your priorities in people's minds during the year, by having them inform the agenda of department meetings, for example. If things change during the year, then review the plan and update it as appropriate. It should remain a working document for you, even if your SLT only requires it to be published at the start of the year.

Developing consistency through implementation of departmental policies

The school effectiveness research also suggests that consistency within the department is an indicator for success, and therefore we recommend the development of departmental policies to provide clarity about the approach to key aspects of your work.

Where the school policy is very explicit, then this may not be necessary. However, there are sometimes differences in interpretation of whole school policy between different subject departments, and making these clear is helpful in bringing about consistency, particularly for less experienced teachers and those who have recently joined the team.

The extent to which science leaders have autonomy to develop their departmental policies varies from school to school. Brown *et al* (2000) suggest that the leadership of the Headteacher and the SLT may be less important than is sometimes thought. Indeed, they go so far as to argue that it is successful middle leadership that underpins the success of the whole school. The authors found that there is variation in the extent to which the interventions of the SLT are found to be helpful by HoDs, with some HoDs feeling that they are the buffer between their science team and the demands of national curricula, school inspectors and the SLT. Some HoDs would like greater autonomy to prioritise what they feel is most important to departmental development. However, departmental autonomy can be less good if change is needed and the department is stuck in a rut.

Where possible, we would recommend that the science department has its own policies to cover the following areas:

➢ Assessment, including marking and report writing (See Chapter 18).

➢ Homework. Departments need to consider how to make homework into meaningful learning opportunities. Douglas (2017) describes how his school uses a flipped learning model at GCSE: students carry out preparatory work before the lesson as homework, then lesson time is spent on application and other more intellectually challenging activities, with teacher support available.

➢ Classroom and behaviour management (see Chapter 8). Managing science classrooms requires development of subject-specific expertise, because, in addition to the challenges of creating a positive learning environment, there is the added challenge of practical work (see Chapter 12). We recognise that adaptations to the whole school policy may not be possible in all schools. One example of an adaptation that may be successful in a science department is that, when students need to leave the classroom, they go to another science class. It may also be helpful to have a departmental detention to share the load between all members of the science team.

➢ Health and safety (See Chapter 24).

➢ Inclusion (see Chapter 9), including student grouping, which is further explained below.

➢ Integrating numeracy and literacy into science lessons (See Chapters 14 and 16).

Student grouping

Disagreement on student grouping was a specific example in the school effectiveness research that might make departments less effective (Sammons et al, 1995). We mention it here because it is often an area of mismatch between the ideas of government ministers, common practices in schools and research findings.

Grouping by ability tends to enjoy the support of politicians. David Cameron is quoted as saying the following whilst he was Shadow Education Secretary, for example:

'Every parent knows that children do best when they are engaged at the right level of ability. So I want to see setting in every single school… Parents know it works. Teachers know it works. Tony Blair promised it in 1997. But it still hasn't happened. We will keep up the pressure until it does' (Cameron, 2006).

Meanwhile, the OECD PISA survey (OECD, 2013) suggests that ability grouping is already more common in the UK than in many other countries (at least in mathematics). The available research evidence tends to suggest that ability grouping produces 'lower...pupil performance overall' (Baines, 2012 p.40).

Whilst it is not our intention to say to science leaders that they are wrong to use ability grouping in their schools (where this is the case), this situation does tend to suggest that further investigation is needed. What is our reasoning for the way in which we group students in our schools? Does ability grouping lead to lower average attainment in our schools than might otherwise be expected? Are some groups of students disadvantaged by ability grouping in our schools (e.g. students from lower socio-economic status backgrounds)? What might we do differently, if needed?

In some schools, science leaders are constrained on this issue by whole school policy, which may itself be more in response to parental expectations than the evidence base. Where students are streamed across several subjects, their groups may not necessarily reflect their attainment in science. There are particular issues for students with EAL (English as an Additional Language) and SEND (Special Educational Needs and Disabilities), who may find themselves in lower-attaining groups because that is where the teaching assistant is deployed. Another difficulty is that it can lead to pigeonholing of teachers – with colleagues missing out on the experience of working with the full range of learners.

Retention of staff

There is significant concern about the recruitment and retention of secondary school teachers in general (NAO, 2017), and science teachers in particular (Allen & Sims, 2017). It is suggested that high staff turnover and rates of attrition amongst new teachers damage students' learning. So, what can we do as science leaders?

A number of factors have recently been reported as potentially important in teacher retention. Teachers' excessive workloads have been identified as a key reason why teachers of all subjects leave the profession (NAO, 2017). We highly recommend that science leaders refer to recent guidance about reducing workload (DfE, 2016).

Allen and Sims (2017) suggest a number of factors specific to science teachers: the availability of high quality subject-specific continuing professional learning opportunities was linked with retention of science teachers. They also cited Donaldson and Johnson (2010), who found that where teachers were teaching mixed subjects they were less likely to be retained, suggesting that science teachers may be disproportionately affected by teaching outside their specialism. They also referred to Smithers and Robinson (2008, in Donald and Johnson, 2010)), who suggested that some physics specialists prefer to teach mathematics than the other sciences.

Some schools adopt strategies that enable teachers to teach the same topic to different groups. Overall, there is a reduction in planning time, even when adapting lessons for different learners. An example is to have three groups, each taught biology, chemistry and physics by specialist teachers, each teacher repeating their lessons with each of the three groups.

In conclusion

It is difficult to cover all aspects of the role in a relatively short book chapter, so we have attempted to take a step back and look at the bigger picture of departmental leadership. We would recommend that you carry out further reading (suggested below) and take up relevant professional learning opportunities.

Readers might consider completing the National Professional Qualification for Middle Leadership (NPQML) (Gov.uk, 2017). STEM Learning Network courses (https://www.stem.org.uk) were recently linked with improved teacher retention (Allen & Sims, 2017). Additionally, many universities offer Masters' courses, which include science education or leadership.

Further reading

ASE (2015) *Science leaders' survival guide* [online resource]. Available at: https://www.ase.org.uk/journals/science-leaders-survival-guide/2015/ Accessed 07.08.17

Fleming, P. (2014) *Successful Middle Leadership in Secondary Schools.* Abingdon: Routledge

Leask, M. & Terrell, I. (2014) *Development Planning and School Improvement for Middle Managers.* Abingdon: Routledge

References

Adey, P. & Shayer, M. (2006) *Really raising standards: Cognitive intervention and academic achievement.* Abingdon: Routledge

Allen, R. & Sims, S. (2017) *Improving Science Teacher Retention: do National STEM Learning Network professional development courses keep science teachers in the classroom?* London: Wellcome Trust. [Online document]. Available at: https://wellcome.ac.uk/sites/default/files/science-teacher-retention.pdf Accessed 16.09.17

Archer, L., Moote, J., Francis, B., DeWitt, J. & Yeomans, L. (2017) 'Stratifying science: a Bourdieusian analysis of student views and experiences of school selective practices in relation to "Triple Science" at KS4 in England', *Research Papers in Education*, **32,** (3), 296–315

Baines, E. (2012) 'Grouping pupils by ability in schools'. In *Bad education: Debunking myths in education*, Adey, P. & Dillon, J. (Eds.), pps. 37–56. London: McGraw-Hill Education

Biggs, J.B. & Collis, K. (1982) *Evaluating the Quality of Learning: The SOLO Taxonomy.* New York: Academic Press

Brown, M., Rutherford, D. & Boyle, B. (2000) 'Leadership for School Improvement: The Role of the Head of Department in UK Secondary Schools', *School Effectiveness and School Improvement*, **11,** (2), 237–258

Cameron, D. (2006) *Cameron promises 'subject streaming' in schools* [Online document]. Available at: http://www.conservativehome.com/thetorydiary/2006/01/cameron_promise.html Accessed 10.08.17

diSessa, A. (2005) 'A history of conceptual change research: Threads and fault lines'. In *The Cambridge Handbook of the Learning Sciences*, Sawyer, R.K. (Ed.), pps. 265–282. Cambridge: Cambridge University Press

DfE (2016) *Reducing teacher workload*, [Online document]. Available at: https://www.gov.uk/government/publications/reducing-teachers-workload/reducing-teachers-workload Accessed 16.09.17

Donaldson, M. & Johnson, S.M. (2010) 'The Price of Misassignment : The role of teaching assignments in Teach For America teachers' exit from low-income schools and the teaching profession', *Educational Evaluation and Policy Analysis*, **32,** (2), 299–323

Douglas, E. (2017) 'More content and more depth: coping with the new GCSEs', *School Science Review*, **98,** (365), 15–19

Driver, R., Rushworth, P., Squires, A. & Wood-Robinson, V. (Eds.) (2014) *Making Sense of Secondary Science: Research into children's ideas (2nd Edition).* Abingdon: Routledge

Dudley, P. (2011) 'Lesson study development in England: from school networks to national policy', *International Journal for Lesson and Learning Studies*, **1,** (1), 85–100

Education Endowment Foundation (2017) *Let's think secondary science,* [Online document]. Available at: https://educationendowmentfoundation.org.uk/our-work/projects/lets-think-secondary-science/ Accessed 06.10.17

Gibb, N. (2015) *Assessment After Levels.* A speech given 25th February 2015. Available at: https://www.gov.uk/government/speeches/assessment-after-levels Accessed 16.09.17

Gov.uk (2017) *National Professional Qualification for Middle Leadership,* [Online document]. Available at: https://www.gov.uk/guidance/national-professional-qualification-for-middle-leadership-npqml Accessed 07.08.17

Gov.uk (2016) *Ofsted inspection: Myths.* [Online document]. Available at: https://www.gov.uk/government/publications/school-inspection-handbook-from-september-2015/ofsted-inspections-mythbusting Accessed 16.09.17

Harlen, W. (Ed.) (2010) *Principles and big ideas of science education,* [Online document]. Available at: https://www.ase.org.uk/resources/big-ideas/ Accessed 08.08.17

Harlen, W. (Ed.) (2015) *Working with the big ideas of science education.* [Online document]. Available at: https://www.ase.org.uk/resources/big-ideas/ Accessed 08.08.17

Institute of Physics (2017) *PIPER (Practical Implications of Physics Education Research).* [Online documents]. Available at: http://www.iop.org/education/teacher/support/piper/page_62597.html Accessed 08.08

Millar, R. & Osborne, J. (Eds.) (1998) *Beyond 2000: Science education for the future: A report with ten recommendations.* London: King's College London, School of Education

National Audit Office (2017) *Retaining and developing the teaching workforce.* [Online document]. Available at: https://www.nao.org.uk/wp-content/uploads/2017/09/Retaining-and-developing-the-teaching-workforce.pdf Accessed 16.09.17

OECD (2013) *Selecting and grouping students.* [Online document]. Available at: https://www.oecd.org/pisa/keyfindings/Vol4Ch2.pdf Accessed 10.08.17

Ofsted (2015) *Key Stage 3: The wasted years?* [Online document]. Available at: https://www.gov.uk/government/publications/key-stage-3-the-wasted-years Accessed 09.08.17

Oswald, S. (2006) 'Leading and managing change in a science department'. In *ASE Guide to Secondary Science Education*, Wood-Robinson, V. (Ed.). Hatfield: Association for Science Education

Sammons, P., Thomas, S. & Mortimore, P. (1995) 'Accounting for variations in academic effectiveness between schools and departments'. In *European Conference on Educational Research*, (pps. 14–17). Bath, UK

Scheerens, J. (1992) *Effective Schooling: Research, theory and practice*. London: Cassell

Smithers, A. & Robinson, P. (2008) *Physics in Schools IV: Supply and Retention of Teachers*. University of Buckingham: Centre for Education and Employment Research

Spielman, A. (2017) *Enriching the fabric of education,* speech given at the Festival of Education, 23rd June 2017. Available at: https://www.gov.uk/government/speeches/amanda-spielmans-speech-at-the-festival-of-education Accessed 07.08.17

CHAPTER 24

Health and safety in science

Steve Jones

This chapter looks at health and safety in the school science laboratory and describes some of the legislation identifying who is responsible for what. It also explains the importance of good practice and gives clear guidance for subject leaders and classroom teachers so that appropriate training, decision-making and recording good practice are easily managed.

Introduction

Teaching and learning science in schools is safe.

However, teaching practical science involves using equipment and materials that are hazardous and which might result in injury if the risks are not properly controlled. Accidents do happen and the consequences are occasionally serious. The best way to avoid these is to ensure that the management of health and safety is a fundamental part of how science departments, teachers and technicians work. In doing this, we will also find that we can prepare for and teach science effectively. Health and safety does not curtail science teaching and can empower it through the confidence of knowing that practical activities, their preparation and use in class, are, indeed, healthy and safe.

The legal framework for safety in school science

The main legislation governing health and safety is the Health and Safety at Work Act 1974 (HSW Act). As well as describing the principles governing health and safety at work, it provides for additional detailed regulations to be enacted by succeeding governments. Examples of these include the Control of Substances Hazardous to Health (COSHH) Regulations; the Management of Health and Safety at Work Regulations; the Pressure Systems Safety Regulations; the Manual Handling Operations Regulations; and many more. Many of these have implications for schools and school science, and school employers and employees are expected to conform to them.

The main purpose of the HSW Act is to protect *employees*. The legislation places duties on the employer to take account of the health and safety of the employees in their work.

Important duties of employers are to provide:

- healthy and safe working conditions for employees;
- healthy and safe conditions for others; and
- a health & safety policy.

Employees, in turn, are also given duties, the main ones being:

- to take reasonable care for their own health & safety and that of other people;
- to co-operate with their employers;
- not to interfere with or misuse equipment provided for health & safety; and
- to inform the employer of any serious defects and failures in health and safety arrangements.

There is, therefore, an overriding requirement for employers to ensure that the work and the working environment are as safe as they can be, and for employees to work safely within that environment. In addition, both employers and employees must take all reasonable care for others who might be affected by or during that work. In schools, students, as well as parents and other visitors, are 'others'.

Because the HSW Act allows for subsequent additions through the introduction and updating of Regulations, it is important that science departments keep up-to-date. This need not be difficult.

ASE, CLEAPSS and the Scottish Schools' Education Research Centre (SSERC) all provide updated information in journals and newsletters. A department needs to ensure that these are read, and there is a mechanism for them to be discussed by all science staff and incorporated into practice as necessary. Remember, each new or updated regulation is intended to improve the overall implementation of the HSW Act in order to build a healthy and safe working environment.

Responsibility for health and safety

Think of responsibility like a chain of command. Responsibility rests at each level of the hierarchy as long as the individual has appropriate experience, expertise and/or training. Responsibility for ensuring this often rests with an individual's line manager. Ultimate responsibility for health and safety (H&S) lies with the employer. In a school, the 'employer' will depend on the type of school as shown below.

England and Wales	
School type	**Employer**
Community schools	The local authority
Community special schools	
Voluntary controlled schools	
Maintained nursery schools	
Pupil referral units	
Foundation schools	The governing body
Foundation special schools	
Voluntary aided schools	
Independent schools	The governing body or proprietor
England	
Academies and free schools	The Academy Trust
Scotland	
State schools funded through Scottish local authorities	The local authority
Independent private or fee-paying schools	The proprietor, board of trustees or equivalent
Grant aided schools	The governing body or equivalent

(Source: HSE FAQs http://www.hse.gov.uk/services/education/faqs.htm#general)

The chain of H&S responsibility in a school can be described as follows:
The principal responsibility for ensuring that working practices are healthy and safe falls to the employer. In order to discharge that responsibility, the employer has to provide healthy and safe working arrangements and ensure that staff are trained to work in a manner that is also healthy and safe.

> ➤ The Headteacher takes overall responsibility for routine management of health and safety within the school, possibly with some active support from one or more governors.
> ➤ Deputy Headteachers take some management responsibility for H&S, depending on their job description and line management functions.
> ➤ Heads of Department (HoDs) generally take responsibility for day-to-day management of H&S within their department.
> ➤ Within a science department, individuals may be given responsibility for part of the science curriculum (such as chemistry, biology or Key Stage 3, age 11-14). This may include responsibility for H&S.
> ➤ Individual teachers are responsible for the H&S aspects of what goes on in their lessons.
> ➤ Technicians are responsible for what they do to support practical activities. In addition, they may support teachers by providing H&S information; however, it is not their responsibility to do the risk assessment for teachers (see later section on risk assessment).

Managing health and safety in a science department

In most schools, one person, generally referred to as the Head of Science, has overall responsibility for managing the department. Within this, explicitly or implicitly, is the management of H&S in science. This means:

> ➤ ensuring that the department structures, practices and equipment conform to all relevant H&S regulations;
> ➤ ensuring that suitable safety guidance, including model risk assessments, is available for staff to consult;
> ➤ ensuring that staff know to follow all appropriate control measures, both in teaching and in the preparation of lessons;
> ➤ monitoring staff practice to ensure that the control measures are in place and are being followed;
> ➤ identifying shortfalls in practice or resources, and reporting these to their line manager; and
> ➤ providing or arranging training as needed.

Department health and safety policy

Probably the most straightforward way to manage H&S in science is to have a well-considered and agreed H&S policy. Some education employers (schools or local authorities) will require the science department to have such a policy, but this is not a specific legal requirement under H&S legislation.

Having an agreed and understood policy means that all staff know what is expected of them in relation to healthy and safe practice, and what they can expect of the department and their colleagues. It provides staff with the confidence that systems and structures are in place to ensure that everyone knows what they can and cannot do, that equipment and materials are safely stored and maintained, and that staff are kept up-to-date.

A suitable policy will include the following:

- General principles;
- Health and safety roles and responsibilities within the department;
- Communication strategies within the department;
- Procedures for monitoring and checking H&S practice within the department;
- An outline of the expectations for staff induction and training;
- Guidance on risk assessment, including sources of model risk assessments and how adaptations to these are recorded in department documentation;
- Guidance on particular equipment and materials, including routine testing and/or inspection;
- Other issues such as secure storage of equipment, and manual handling;
- Emergency procedures;
- Laboratory rules for pupils; and
- Staff roles and emergency contacts.

CLEAPSS and SSERC offer their members a model policy, which will include all of the above and which can be customised by an individual department. These make the physical production of a department H&S policy quite easy, but Heads of Science should recognise that the policy itself is of little value if it is not fully understood, agreed and implemented by all science staff.

Structures and practices within the department

As previously mentioned, the Head of Science is responsible for providing and maintaining effective structures and practices to ensure H&S. These are easily identified from a sensible H&S policy. However, issues rarely arise from a lack of structures or policies; more often it is the failure of individuals to implement them that leads to problems. This often stems from a lack of real agreement about details in the H&S policy. It is essential that this is discussed and understood from the outset, so that all staff understand and agree with the policy. Another frequently articulated issue is that of insufficient funding for appropriate H&S equipment or equipment to ensure that practice remains healthy and safe: 'We don't have the money for ...'. The only point to note here is that a lack of funding is never taken as an adequate reason for a failure to maintain reasonable and appropriate H&S standards and practice.

Monitoring health and safety

Although a HoD may have the responsibility for H&S in respect of the preparation for and teaching of science, the implementation of this rests with everyone else in the department. The Head of Science has the responsibility to check that practice accords with policy and therefore needs a planned programme to monitor practice. A useful element of this would be to include H&S as a standing item on the agenda of all department meetings. Discussion can be as full or as brief as is needed, but it is important that one or more technicians are included in the meeting for this part of the agenda. That way, it is possible to routinely identify any areas of concern in the department, which might subsequently be the focus of closer scrutiny. The HoD may not personally need to undertake that closer scrutiny but can delegate the task, possibly as part of professional development, to other staff who report back at a subsequent meeting.

Induction and training

All new staff, and those whose job changes significantly, must be given training in the H&S aspects of their work. This is an employer's duty under the Management of Health and Safety Regulations. In schools, the task of ensuring that this happens would fall, explicitly or implicitly, to the Head of Science. Newly qualified teachers, those new to the department, any given additional responsibility and any trainee teachers all need some form of induction and training. Induction is necessarily a department task, but training may be provided by outside organisations as well as more experienced teachers or technicians within the department.

The issue is not who provides training, but that it *is* provided, and that it meets the identified needs of the individual. Publications from ASE, CLEAPSS and SSERC can be used to help identify needs and construct a programme to meet them.

It is often assumed that a well-qualified teacher or technician can arrange and manage their own induction and training. They can read department publications and ask pertinent questions as required. Indeed, they are encouraged to ask whenever they need to know something. Often, though, they, especially new or trainee teachers and technicians, don't know what they might need to ask about until they discover in the middle of a lesson or task that they do not know fully what they are doing. This situation should not be allowed to arise and will not where a department has a sensible induction and training programme in place. It is true that preparing and helping to deliver and monitor such a programme takes time, often from several staff. Under the Management of Health & Safety Regulations, this time has to be found, because there is no avoiding the employer's duty. A Head of Science should clearly identify this to the school senior leaders whenever recruitment is being arranged.

Risk assessment

The distinction between a 'hazard' and a 'risk' is a cornerstone of managing safety in the workplace in the UK. Whilst these two terms are often used interchangeably in everyday language, the distinction is useful in this context. Essentially, a hazard is an intrinsic property of the material or process that cannot be changed, whilst the risk is related to the likelihood that a given hazard will cause significant harm.

The focus in UK legislation is on risk as opposed to hazard. For example, concentrated sulphuric acid is corrosive – however, what matters is whether it is likely to cause significant harm if used in a particular way. Unlike hazard, risk is activity-specific and can be reduced through the use of suitable control measures – for example, in this case using lower concentrations, smaller quantities and by wearing appropriate Personal Protective Equipment (PPE).

The requirement for 'risk assessment' was first described in the first edition of the COSSH Regulations in 1988. Prior to this, legislation and regulations often consisted of rules, which had to be followed, under threat of prosecution for failing to do so. The principle of risk assessment gives the employer the responsibility to decide for his/her work the risks to the H&S of the employees, and others, and how these can be reduced. There is a responsibility to reduce the risk to the lowest level that is 'reasonably practicable', and prosecution can

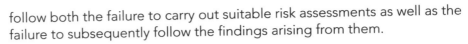

follow both the failure to carry out suitable risk assessments as well as the failure to subsequently follow the findings arising from them.

The theoretical process of risk assessment is straightforward:

- ➤ Identify the hazards;
- ➤ Assess the risk from those hazards, i.e. how likely is it that this hazard will cause injury;
- ➤ Identify the severity of any injury, i.e. how many individuals are likely to be affected, and how severe might be the injury; and
- ➤ Take steps to control the risk (control measures) and reduce it to as low a level as reasonably practicable.

In practice, the process is often complex for school science as it is virtually impossible for an employer to carry out formal risk assessment for every activity that might be undertaken.

The Management of Health and Safety at Work Regulations (1999, n.p.), however, allow the use of *model* risk assessments, which can be provided by the employer or by other organisations. The guidance that accompanied these Management Regulations included the following statements:

'Employers who control a number of similar workplaces containing similar activities may produce a basic "model" risk assessment reflecting the core hazards and risks associated with these activities. "Model" assessments may be developed by trade associations, employers' bodies or other organisations concerned with a particular activity. Such "model" assessments may be applied by employers or managers at each workplace, but only if they:

(a) satisfy themselves that the "model" assessment is broadly appropriate to their work; and

(b) adapt the "model" to the detail of their own actual work situations, including any extension necessary to cover hazards and risks not referred to in the "model"'.

Organisations such as CLEAPSS and, in Scotland, SSERC provide model risk assessments, in a variety of formats, in their publications. The majority of education employers in the UK have instructed their employees to use these to manage the risks arising from preparing for and teaching practical science.

What do science teachers and technicians need to do?

The process of using and adapting a model risk assessment is as follows:

➢ Make sure you have access to the full range of model risk assessments relevant to your work;

➢ Consult the model risk assessment advice for the activity or activities that you are planning;

➢ Consider whether the advice in the model risk assessment fully or partially meets your particular circumstances and needs. Common points to consider include:

   ○ class size – is your class larger or smaller than average?
   ○ students' experience and expertise, is it average? Will the students be able to follow the instructions and handle the equipment and materials correctly?
   ○ will pupils behave appropriately?
   ○ are there particular features about your room layout or size that might impact on the risks?
   ○ do you have sufficient experience or expertise to manage the class and the activity?
   ○ is it better done as a demonstration rather than a class practical?;

➢ Once you have decided these issues, the Regulations require that you make a note of any *significant* adaptations in a *'point of use text'*. The Regulations do not specify what the 'point of use text' is, beyond the fact that it should be available to the person carrying out the activity at the point at which they are actually planning or carrying it out. For this reason, a form, labelled 'risk assessment', which is kept only in a file or in an electronic folder on a shared area is unlikely to count as a 'point of use text'; and

➢ Then, follow the outcomes of your adapted risk assessment.

Making the risk assessment process manageable

Two aspects are central to making the risk assessment process manageable:

➢ Working as a team across the department to adapt the practical activities and to customise the advice contained in the model risk assessment to local circumstances. Without a team approach, this is likely to be unmanageable for individual teachers; and

➢ Identifying suitable 'point of use texts' that are already in use in the department in which to record the significant outcomes from the above. This documentation should be easily accessible to all staff and be central to planning, supporting and delivering practical lessons. They are likely to be different for teachers and for technicians.

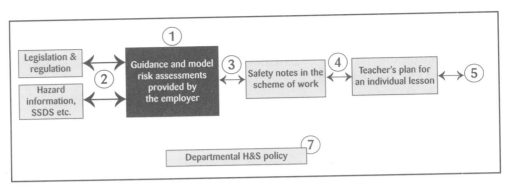

A departmental approach to using model risk assessment *for teachers* to plan and deliver safe practical lessons

1. The employer provides access to model risk assessment advice and guidance for the activities the department wishes to carry out.

   The organisation generating the model risk assessments will have considered all the hazard information (for example, hazard classifications for chemicals provided in suppliers' safety data sheets (SSDS)) and any relevant legislation, and interpreted this in a proportionate way appropriate to a generic school context. The overwhelming majority of educational employers use materials provided by CLEAPSS (in England, Wales and Northern Ireland) and SSERC (in Scotland).

2. As a team, the teachers and others in the department consider each practical activity and consult the relevant guidance included in the model risk assessment, adapting each activity as necessary to bring it in line with the advice.

3. The outcomes from this process are recorded in the scheme of work (or other 'point of use text') and key safety points are added as 'safety notes' for all teaching staff to use.

4. Teachers refer to the adapted scheme of work when planning their individual lessons. They take into account any specifics relating to their class and make a note of these on whatever documentation they use to plan their lessons – for example, on a lesson planning form, in their Planner or on a requisition sheet.

5. Teachers refer to these lessons-specific adaptations when they teach the lesson. Remember that risk assessment is a dynamic process, which, for a given lesson, is not finished until that lesson is over.

6. Where teachers intend to use new activities not currently included within the scheme of work, they should be aware that they will need to spend a bit longer checking these against model risk assessment guidance, as this will not have been done for them by colleagues. It would be good practice to implement a system whereby any new activity – particularly one that is novel in some way – should be signed off by the HoD before being attempted in a classroom.

7. These processes and the expectations for staff are explained in a section on risk assessment in the departmental handbook or safety policy.

Possible point of use texts
For teachers:
➢ Scheme of work

➢ Lesson plans

➢ Teacher's planner

For teachers and/or technicians:
➢ Requisition sheets

➢ Worksheets/practical sheets (from the scheme of work)

For technicians:
➢ Technicians' notes sections/pages from a scheme of work

➢ Recipe sheet/ Hazcard

➢ Technicians' notebook

Teaching health and safety
A relatively new development in the school curriculum has been a requirement for students to learn not just how to follow instructions and work safely, but also something about the principles and practice of risk assessment itself, the intention being for students to develop an awareness of the need to manage risks both in the workplace and in their everyday lives.

Science offers many opportunities to teach and practice these ideas. For any practical activity, including investigations, students can be invited to consider the risks, and put forward their own suggestions on how to manage these. It is a requirement of some science examination programmes, including up to A-level and equivalent, that they do this as part of an individual project or investigation. Teachers should remember, though, that the responsibility for healthy and safe practice in those lessons and activities rests with the teacher. It is essential that the teacher checks and verifies any risk assessment and control measures that students produce. Teachers must use their judgement to balance the requirement to teach and support students' developing understanding with the need for effective H&S.

Useful websites

Association for Science Education (ASE): www.ase.org.uk

Health and Safety Executive (HSE):
www.hse.gov.uk/services/education/index.htm

Scottish Schools Education Research Centre (SSERC): www.sserc.org.uk

The school science and technology safety organisation – CLEAPSS:
www.cleapss.org.uk

References

The Management of Health and Safety at Work Regulations 1999
Printed and published in the UK by The Stationery Office
WO 5858 12/99 466785 19585
ISBN 0-11-085625-2
http://www.legislation.gov.uk/uksi/1999/3242/contents/made

CLEAPSS (2005) *L196 Managing Risk Assessment in Science*

CLEAPSS (2017) *PS090 Making and recording risk assessments in school science*

CONCLUSION

We very much hope that you have enjoyed reading this book, that it has encouraged you in your career as a science teacher, and that it has made you think about your practice in the classroom and how you might develop it further, drawing on some of the evidence presented by the contributors to this book.

One of the hallmarks of being part of the science teaching profession is being a reflective practitioner, and using evidence to inform that process. As teachers, you can use both the evidence gathered by others, and evidence that you collect yourself. You also need to think about how to interpret that evidence, and the forthcoming *ASE Guide to Research in Science Education* will help with both the collection and interpretation of this evidence. You are welcome to share your findings with others, through the various ASE publications, *Education in Science* and *School Science Review*, and at the ASE regional and national conferences.

We suggest that you keep this book and refer back to it, that you discuss some of the issues we have raised with your colleagues and that you look up some of the further reading listed at the end of each chapter. You may even recommend this book to other science teachers! But, most importantly, we would encourage you to become part of the science teaching community that is the Association for Science Education, the publisher of this book, and join with us as we work together to promote excellence in science teaching and learning.

Indira Banner and **Judith Hillier, Editors**
University of Leeds and University of Oxford
July 2018

INDEX

Numbers relate to chapters

Active learning 7
Assessment 1, 4, 7, 18, 19, 20, 23
Assessment for Learning/AfL 18, 19
Assessment, summative 18, 19

Beginning teachers/ NQTs 21, 22
Behaviour management 8
Belief (s) 2

Careers 5, 6
Classroom management 23, 24
Collaborative/ co-operative learning
    5, 6, 13, 15, 19
Constructivism/ constructivist
    1, 16
Continuing Professional Development
    (CPD) 21
Creativity 11
Critical thinking 5
Cross-curricular 11, 14
Cross-phase/transition 4, 23

Data in schools 19
Dialogue/ dialogic teaching 16
Differentiation 6
Discrimination 9, 10
Diversity 9, 10
Double Award Science 23
Drama 11

EAL (English as an additional language)
    9, 10
Emotion(s) 3, 5
Equality 9, 10
Ethics/ values 2
Ethnicity 10

Formative assessment 18, 19

Gender 10
Gifted 2

Health and safety 4, 9, 24

ICT 15
Inclusion 9
Inequality 9
Inquiry/enquiry 1, 7, 11
International (assessment) 20

Knowledge Quartet 8

Language/literacy 9, 14, 15, 16, 17
Leadership 23
Levels 19
Literacy, scientific 1, 9, 16, 17

Mathematics in science 14
Mentor/mentoring 22
Mixed ability 23
Models 7
Motivation/ motivational theory 3, 6

NQT/ beginning teachers 21, 22
Nature of science 2
Neuroscience 3
News, Science in the 11, 17
Numeracy 14

Outdoors/ outdoor learning 13

PCK (Pedagogical Content Knowledge)
     8, 22
Passive learning 7, 18
Planning 6, 8, 12, 13, 22, 23
Practical work in science 12, 22, 24
Primary science 4
Professional development 21

Questions/questioning 11, 13

Reasoning 3, 17
Recording 19
Reflection 21
Religion/ beliefs 2
Resilience 21
Retention (teacher supply) 21, 23
Risk, risk assessment 24

SEND 9, 10
Scaffolding 8, 12
Science Capital 6, 10
Science in Society 6
Social media 15
Streaming 23
Summative assessment 18, 19

Talk/talking 7, 16, 17
Teacher supply/retention 21, 23
Thinking/thinking scientifically
     2, 16, 17
Transition/transfer 23
Triple Science 23

Values/ ethics 2

Wellbeing 21